Architectural Development of China’s Foreign-aided Stadiums: 1956-2019

中国援外体育场馆建筑创作历程：1956—2019

常 威 著
Chang Wei

图书在版编目(CIP)数据

中国援外体育场馆建筑创作历程 : 1956-2019 = Architectural Development of China’s Foreign-aided Stadiums : 1956-2019 : 英文 / 常威著. -- 天津 : 天津大学出版社, 2024.2

ISBN 978-7-5618-7675-6

Ⅰ. ①中… Ⅱ. ①常… Ⅲ. ①体育场－建筑设计－建筑史－中国－1956-2019－英文②体育馆－建筑设计－建筑史－中国－1956-2019－英文 Ⅳ. ①TU245

中国国家版本馆CIP数据核字(2024)第046474号

出版发行 天津大学出版社
地　　址 天津市卫津路92号天津大学内（邮编:300072）
电　　话 发行部:022-27403647
网　　址 www.tjupress.com.cn
印　　刷 北京盛通印刷股份有限公司
经　　销 全国各地新华书店
开　　本 787mm×1092mm　1/16
印　　张 12
字　　数 389千
版　　次 2024年2月第1版
印　　次 2024年2月第1次
定　　价 89.00元

Acknowlegement

This book is based on the PhD thesis of the author and is part of the study supported by Humanities and Social Sciences Research Project of Colleges and Universities in Hebei Province (河北省高等学校人文社会科学研究项目: BJS2022039) and Doctoral Fund of Tangshan Normal University (唐山师范学院博士基金项目: 2022A04). This book is sponsored by Tangshan Normal University (该书由唐山师范学院学术著作出版资助).

About the Author

Chang Wei, PhD, member of the Jiusan Society, vice professor of School of Fine Arts in Tangshan Normal University, master supervisor of the architecture major from School of Civil Engineering and Mechanics in Yanshan University, China's national registered urban and rural planner, and member of the "Expert on Urban Renewal of Beijing, Tianjin and Hebei". She got her PhD degree from the City University of Hong Kong majoring in Architecture and Civil Engineering, got her Master of Architecture degree from the University College London and her Bachelor of Architecture degree from Tianjin University. She has been engaged in the teaching and research works about architectural design for years. As the first author, she has published more than 20 academic papers in high-level journals such as A&HCI and SCI. She has also published one academic monograph as the first author and one academic translation book as the sole author. She has conducted and participated in over 10 teaching and research projects at all levels.

Preface

This book investigates the historical development of modernist architectural design in China's foreign-aided stadiums and how Chinese architects designed these cross-border projects in various historical periods.

China's foreign-aided stadium projects occupy a considerable proportion of China's foreign-aided constructions overseas and stand out from other building types of China's foreign-aided constructions due to their unique qualities. These megastructures are located in diversified geographical and cultural contexts, and their designs tend to bc differentiated from Chinese domestic sports buildings to some extent. Such manifestation is believed to be the historical choice and is termed as unique "architectural exportation" in this book. Also, Chinese architects' cleverness, persistence, concession, momentum and breakthroughs while working under limitations have created new attributes to the exportation.

Designed by Chinese architects from China's state-owned institutions, these foreign aid stadiums are indivisible parts of Chinese architecture and Chinese sports buildings, regarded as the architectural exportation from China to the developing world towards modernization. Unlike China's architectures, the environment and contexts of which are more familiar to Chinese architects, the foreign aid stadiums provide opportunities and challenges for Chinese architects to design with different styles and characteristics from those in domestic stadiums. The characteristics of these buildings have varied over time and across regions, reflecting the development of Chinese architecture, Chinese sports buildings, and the designs of Chinese architects. In terms of the design practice of megastructure architecture, such as the stadiums, the theoretical practice has been characterized by the coexistence of internationalized mainstream of modernity and diversified patterns.

The current body of this research includes three periods, the 1950s–1970s, the 1980s–1990s, and the 21st century, which witnessed the development of the designs of China's foreign-aided stadiums. This research aims to explore the (relatively) decisive influences and main design principles that Chinese architects preferred to adopt in different historical periods. The author has extensively read the related books, articles, and government documents, interviewed the relevant institutes/designers, and visited the targeted buildings on site. Through these endeavours, this book reveals the architectural essence of China's foreign-aided constructions and presents a panorama of Chinese-designed stadiums overseas in the past 60 years.

The book is organized into five chapters. Chapter 1 presents the background, which starts with the background of China's foreign aid, China's foreign-aided constructions, and China's foreign-aided stadiums. Chapter 2, Chapter 3 and Chapter 4, as the backbone of the book, present the historical survey of China's foreign-aided stadiums in three different periods, i.e. the 1950s to 1970s, the 1980s to 1990s and the 21st century respectively. Chapter 3 carries out the specific study of China's foreign-aided stadiums of the initial period, from the 1950s to 1970s. The starting of China's gift stadiums is narrated, and the designs of these stadiums of the early period are analysed, as well as the interviews with Chinese architects. Detailed studies of three cases are also conducted before the final analysis of the impact factors in the summary of this chapter. Chapter 4 contextualizes the features of China's foreign-aided stadiums in the developing period from the 1980s to 1990s. Based on the studies of two representative stadium cases, what remained and what changed in the designs of these stadiums of the second period are analyzed and concluded. Chapter 5 highlights the development characteristics of the designs of China's foreign-aided stadiums of the 21st century. Five cases are particularly illustrated to achieve more comprehensive conclusions about the impact factors in the designs. Chapter 6 is the conclusion part, which concludes the research with discussion from three perspectives. The first discusses the similarities and dissimilarities between China's foreign-aided stadiums and its domestic ones through comparisons. The second reveals the regional design approaches utilized by Chinese architects in various periods and their development trends. The third ends with foreseeing of the future designs, the conclusions of the research findings and contributions, and the limitations and directions for future research.

It is hoped that this book demonstrates the historical vicissitude of China's foreign-aided stadiums, reveals the challenges and opportunities for Chinese architects, provides a reference value for the designs of future cross-border constructions, and further situates modern and contemporary Chinese (sports) architecture within a global context.

CONTENTS

CHAPTER 1
INTRODUCTION

1.1 China's Foreign Aid

Foreign aid refers to a voluntary transfer of public resources from a government to another independent government or other international organizations (Lancaster, 2007). It includes assistance on various aspects such as economy, materials, technology, health care, education, training, etc., from donor countries (regions) to recipient countries (regions). As the international transfer of capital, goods, constructions, or services from a country or international organization for the benefit of the recipient country or its population, foreign aid has become one of the most significant international activities.

China began to assist neighbouring developing countries in the late 1950s (The State Council of China, 2011). Technically, China's foreign aid is defined by its concessional nature and includes grants, zero-interest loans and "concessional" (low, fixed interest) loans, managed by the Ministry of Commerce of the People's Republic of China (MOFCOM), as listed in the Chinese government's White Paper, with eight forms of aid: "complete projects; goods and materials; technical cooperation; human resource development cooperation; medical teams sent abroad; emergency humanitarian aid (disaster relief); volunteer programs in foreign countries; and debt relief" (The State Council of China, 2011). This definition is different from ODA's (Official Development Assistance)[①], which has been standardized for the 24 members of the DAC (Development Assistance Committee); while China is one of the largest emerging non-DAC[②] countries that provide foreign aid.

1.2 China's Foreign-aided Construction

Complete projects are one of China's main aid patterns, accounting for approximately 40% of China's financial expenditure on foreign aid. It refers to projects financed by China's free donations, interest-free loans, low-interest loans or other special aid funds to assist the recipient countries. According to the Ministry of Commerce of the People's Republic of

---

① ODA is a term coined by the Development Assistance Committee (DAC) of the Organization for Economic Co-operation and Development (OECD) to measure aid. ODA's definition of foreign aid is: flows of official financing administered with the promotion of the economic development and welfare of developing countries as the main objective, which are concessional in character with a grant element of at least 25 per cent (using a fixed 10 per cent rate of discount).

② As one of the dominant aid organizations, DAC (Development Assistance Committee) consists of 30 members including U.S., U.K., Japan, Korea and some European countries. With the development of emerging countries, some non-DAC countries such as China and India contribute considerable proportions of the international aid.

China (MOFCOM), till 2015, China had exported over 2,700 complete projects (anonymous, 2015). Basically, most of China's foreign aided constructions are carried out in the form of complete projects. In recent years, China has focused more on the local infrastructure in fields such as education, sports, residents, municipal facilities and other public buildings, which may benefit more to the recipient conntries.

From 1960 to 2009, China constructed numerous complete projects in over 160 countries, together with other forms of assistance in the areas of health, manufacturing, agriculture, education and sports, totaling RMB 256.29 billion yuan in investment (The State Council of China, 2011). According to the author's on-going database[①], the number of China's foreign-aided constructions (civil) reached 1,500, including government offices, parliament buildings, convention and exhibition centers, stadiums, theatres, schools, hospitals, libraries, railways and railway stations, etc.

1.3 China's Foreign-aided Stadiums

Of China's foreign-aided public facilities projects, sports facilities constitute a considerable proportion (The State Council of China, 2011) (Fig. 1.1), together with equipment and technical assistance (Yu & Yuan, 2010). Over 80% of these sports facilities are stadiums[②] according to the author's statistics. Over 100 (111 in total) China' foreign-aided stadiums had been financed, designed and constructed in 66 countries as of 2019[③]. And the number steadily increased, before the 1980s, then increased more markedly afterwards. Two peak periods appeared around 1987 and 2007, when China experienced its tides of sports development promoted by 1990 Asian Games and 2008 Olympic Games held in Beijing, respectively. The total number surpassed 100 after the 21st century, and seems set to increase further under the BRI (Belt and Road Initiative) (Fig. 1.2). These stadiums take approximately 60% of all China's foreign-aided public buildings, being the building category that accounts for the largest number of China's foreign-aided constructions. The type of these stadiums includes outdoor and indoor stadiums with various specialties such as football, basketball, gymnastics, swimming and diving, cricket, etc. (Appendix 2)

① This number is from the author's own collected database. Sources include government and design/construction companies' websites and news reports.

② In this study, the China's foreign-aided stadiums refer to the public stadiums of the local, mainly including outdoor and indoor stadiums. Sports facilities of the schools or universities are not included.

③ The number of China's foreign-aided stadiums was calculated by the author.

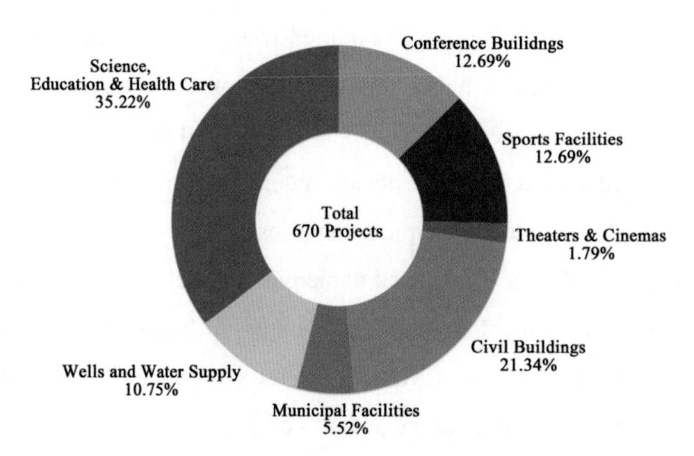

Fig. 1.1　Sectorial distribution of complete overseas public facilities projects constructed with the help of China before 2009 (Source: drawn by the author, based on the information from "China's White Paper on Foreign Affairs, 2011")

These stadium projects were initially implemented in Asia and Africa (the 1960s to 1970s), then spread to Oceania and Latin America (the 1980s-1990s), and even to Europe (after 2000). Still, the highest concentration was in Africa (Fig. 1.3). The wide range of regional distribution leads to disparate geographies, climates, diversified levels of urban development, human histories and other complex circumstances facing these megaprojects.

Most of China's foreign-aided stadiums were donated for preparing significant international sports events, and the commitment by the Chinese government was made at the request of the recipient countries. Some others were for the improvement of local sports facilities or other reasons (Table. 1.1). In addition, China has continued to support the maintenance or upgrading of some of its early export sports buildings, upon the requests from the recipient countries sometimes, in case of losing the shine and value in the local community, since some were poorly maintained locally before or left obsolete due to various causes (Appendix 3).

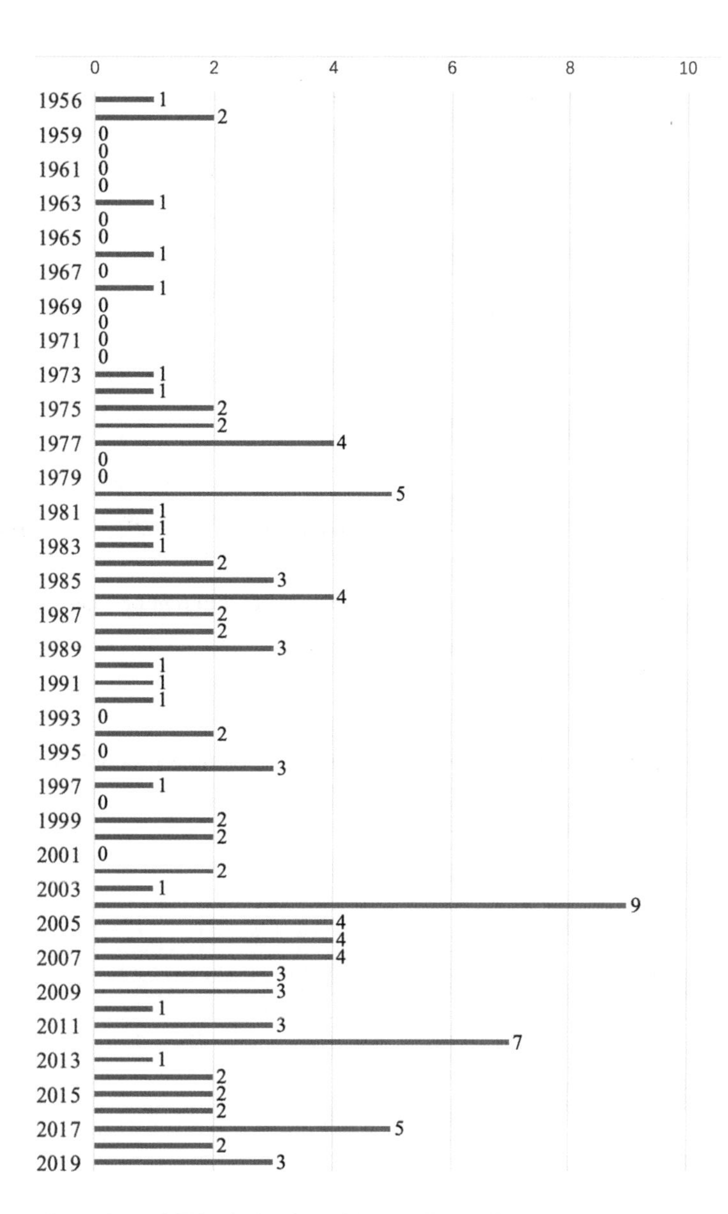

Fig. 1.2 Annual number of China's foreign-aided stadiums (Source: drawn by the author based on the database of this study)

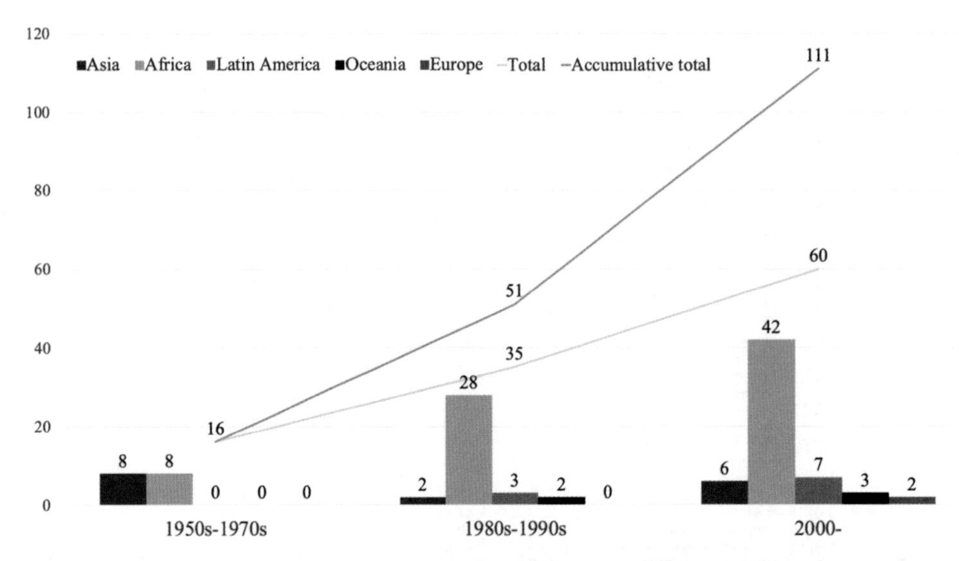

Fig. 1.3 Numbers of China's foreign-aided stadiums by area in different periods (Source: drawn by the author based on the database of this study)

Table 1.1 Purpose of China's foreign-aided stadiums (examples)

Year	Country	Building	Event
1958	Mongolia	National Olympic Stadium	Improvement of the sports facilities
1963	Indonesia	Gelora Bung Karno Stadium (National Stadium)	1st GANEFO
1966	Cambodia	National Sports Complex (and International Village)	2nd GANEFO
1984	Pakistan	Pakistan Sports Complex (Jinnah Sport Stadium)	8th Asian Games (cancelled)
1987	Kenya	Moi International Sports Center	4th All-Africa Games
1997	Barbados	Wildey Gymnasium	Improvement of the sports facilities; regional development
2003	Vietnam	My Dinh National Stadium	22nd Southeast Asian Games
2010	Mongolia	National Gymnasium	Improvement of the sports facilities
2010	Sri Lanka	Mahinda Rajapaksa International Cricket Stadium	10th Cricket World Cup
2014	Ivory Coast	New Olympic Stadium of Abidjan	2021 Cup of African Nations (CAN)
2018	Cambodia	National Stadium	32nd Southeast Asian Games

CHAPTER 2

THE 1950s TO 1970s — INITIAL ASSISTANCE FROM LABOUR TO DESIGN

2.1 Diplomatic Help and Direct Supervision

China began to provide economic and technological assistance to other countries in the 1950s (The State Council of China, 2011). In the 1955 Asia-Africa Conference held in Bandung, Indonesia, the Chinese leader, Zhou Enlai, made the acquaintance of Prince Sihanouk of Cambodia. In delivering his address to the General Assembly, and stating China's position on the foreign aid, the relationships among China and developing countries in Asia and Africa entered a new historical period:

"*We called for the independent economic development of the Asian and African countries. Asian and African countries called for economic and cultural cooperation in order to contribute to the elimination of our economic and cultural backwardness caused by long-term colonial plunder and oppression. Cooperation among Asian and African countries should be based on the foundation of equality and mutual benefit, and no privileged conditions should be attached. Our trade contacts and economic cooperation should promote independent economic development, rather than making any country the source of raw materials and a sales market for consumer goods.*" (The Ministry of Foreign Affairs of China, 1958).

During the early period, the aid projects were directly carried out by Chinese central government systematically. They were generally managed by the Ministry of Economic Relations with Foreign Countries[①] (MERFC). In addition, China established one specific mechanism named "ministry-responsible overall delivery mode", which forced other ministries to participate and share the responsibilities in the management and delivery of aid projects according to their attributes. For example, the assistant industrial materials were in the charge of China's Ministry of Industry, while the textile resource aid by the Ministry of Textiles. Similarly, construction aid projects were managed by the MERFC with the participation of other ministries such as the Ministry of Construction and the Ministry of Communications[②]. The MERFC conveyed the instructions from the central government to the specific bureau[③] in the Ministry of Construction, which took the direct responsibilities for

① The MERFC was reformed to be one independent ministry in 1970 from the Bureau of Economic Relations with Foreign Countries affiliated the State Council of China.

② The Ministry of Communications mainly took the mission of transporting materials and labors for China's foreign-aided constructions.

③ The bureau in the Ministry of Construction was established in 1959 to deal with the aid construction projects.

exporting construction aid overseas.

From the 1950s to 1970s, China-aided construction projects were run in the planned economy model, in which the government designated the institutes responsible for the design and management of the project. China's Ministry of Construction was in charge of investigation, design, material supply, construction, technique support and management process of most aid construction projects. The design work of all construction projects was supposed to be completed by one of the six design institutes① affiliated to the Ministry of Construction — Beijing Industrial and Architectural Design Institute (BIADI). As the first designated institute② of the design of China-aided construction projects overseas, BIADI consisted of four departments: one for the design of China's domestic industrial and civil construction, one for the design of overseas civil aid construction, one for the design of overseas industrial aid construction, and one for the setting up of industrial criteria and standards of the Peoople's Republic of China③ (Fig.2.1).

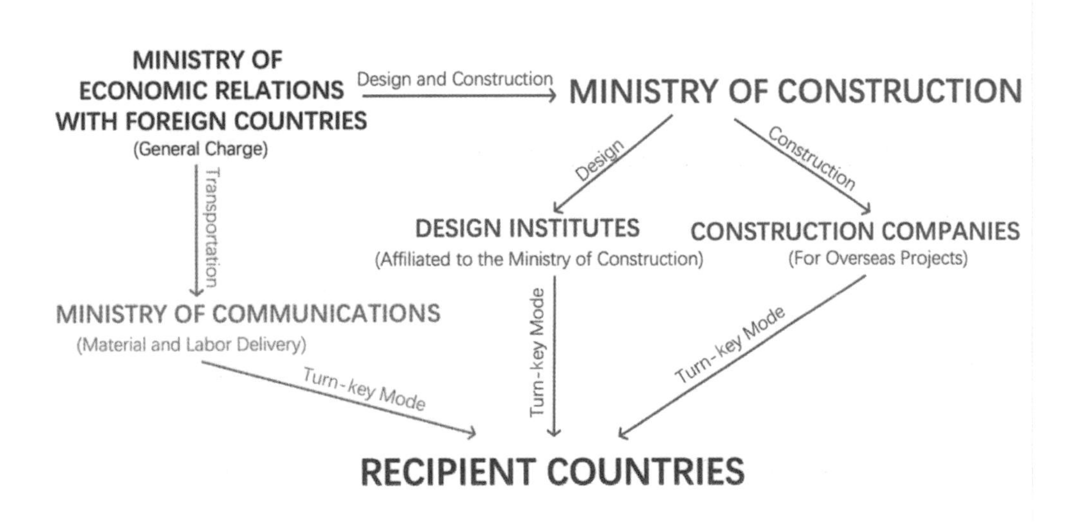

Fig. 2.1 Mechanism of China-aided construction projects from the 1950s to 1970s

① The other five institutes include Water System Design Institute, Infrastructural Design Institute, Planning Design Institute, Electric Power Design Institute and Structural Design Institute.

② Later some of the aid projects were designed by other design institutes located in other cities of China, which were also affiliated to the Ministry of Construction.

③ The fourth department was developed to be China Institute of Building Standard Design & Research in the 21st century, which set the majorities of standards for the Chinese construction industry. BIADI was developed to be China Architecture Design & Research Group (CADRG), as the most core design group of China.

The state-own design institutes received instructions directly from the central government and normally assigned the design work to their leading technicians or managers, such as directors of the institutes, chief superintendents, chief architects and chief engineers. And the projects were handed to the recipient countries through Turn-key mode after construction and exported in the form of "complete projects" by the Complete Plant Import & Export Corporation (COMPLANT①). Therefore, the mechanism of the early period is relatively simple but absolute that all construction aid projects were under the direct control of the central government (Table 2.1).

Table 2.1 Policy and mechanism of China-aided construction projects from the 1950s to 1970s

Period	The 1950s-1970s
Aid mode	Ministry-responsible overall delivery mode
Principles	Eight principles
Management department	Ministry of Economic Relations with Foreign Countries(MERFC), Ministry of Construction
Design institute	State-owned design institute affiliated to the Ministry of Construction (mainly BIADI)
Construction aid mode	Turn-key mode

2.2 Chinese Stadiums from the 1950s to 1970s

Since the establishment of China's new government in 1949, Chinese stadiums experienced development with numbers of them constructed over the land. In this section, this study tries to illustrate it in the 1950s. Later the structural development of Chinese stadiums of the period (1950s-1970s) will be demonstrated, followed by the summary of this section.

2.2.1 Stadium Development under the Socialist Influence in the 1950s

The stadium development started in the early 1950s, after the initial years of the establishment of PRC, during which the new government was busy with the recovery of post-war reconstructions, and little attention was focused on sports buildings, most being on industrial and residential ones. China's participation of Helsinki Olympic Games in 1952, which was the first time that China attended the Olympic Games after its establishment, awakened the Chinese government with the direct relationship between the level of sports

① COMPLANT is a national funded company established in 1959 to export large-scale sports construction projects to recipient countries (Su, 1989).

development and international standing. Therefore, in its "First Five-year Plan" (1953-1957), the construction of stadiums was required specifically under the principle of "emphasized focus", which was to initially meet the needs of citizens' sports activities (Zhang et al., 1984). During this period, 15 large indoor stadiums were built, eight of which had capacities of around 4,000 to 8,000 seats (Ge, 1957); these indoor stadiums were located in major Chinese cities, such as Beijing, Tianjin, Shanghai, Wuhan, Chongqing, Guangzhou, Changchun, Xi'an, Shenyang, Harbin, Hefei and Nanchang. Moreover, several large-scale outdoor stadiums were first built in the metropolis of China, with 30,000 to 50,000 seats.

Beijing, as the capital of PRC, was the first city that started the construction of stadiums, of which the Beijing Stadium became one representative in the 1950s. Designed by Yang Xiliu from Beijing Institute of Architectural Design (BIAD), it cost RMB 7 million yuan, with a gross floor area of 33,700 m^2, a 22.4 m × 36 m-size indoor playground and a capacity of 6,000 seats. It is composed parallel of three main sports space, with the competition hall in the middle, and the practice hall and the swimming hall on the east and west sides. The structure adopted frame and brick-concrete structures. Its construction was completed in 1955 under the instruction of He Long, one of the core leaders of PRC, who was in charge of China's Ministry of Sports at the time.

During the first two decades, the major aspects of Chinese society were influenced by the Soviet Union, with architecture included. In the early 1950s, the Soviet Union began to export large-scale construction aids to China, and Chinese architecture and architects were deeply influenced by the Soviet ideology in their architectural creation. "Socialist in content, national in form"① and "socialist realism"② became the highest guiding principles in architecture. "Palace style", as one of the commonly used categories of "national forms"③, appeared to be the thick coverage of modern mega-structures in the design of Beijing Indoor Stadium. The main entrance absorbed simplified patterns and elements of Chinese tra-

① It was firstly proposed by the Soviet Union, which required the arts and architecture to contain socialist ideologies in the design, with the national forms.

② "Socialist realism" was the predominant form of approved art in the Soviet Union from its development in the early 1920s to its eventual fall from official status beginning in the late 1960s until the breakup of the Soviet Union in 1991 (Korin, 1971).

③ At the time, the "national forms" in Chinese architecture mainly contained two types. One is the Chinese forms, including the basic "palace form" which explored the modern expressions of Chinese traditional palace buildings with new functions, techniques and materials, and the "regional form" which tried to use symbols or elements from the local traditional buildings of various regions in China. The other is the "foreign forms" which imported the Western classic building styles, especially those from the Soviet Unions.

ditional palaces and the four sides' symmetrical facades with traditional decorations, windows and roofs contributed to the holistic feeling of "Chinese nationality"[①](Fig. 2.2). Similarities can be found in other Chinese stadiums of the same period, such as Chongqing People's Stadium[②] (1954) and Tianjin People's Indoor Stadium (1956) (Fig 2.3).

Fig. 2.2 Beijing Indoor Stadium (1955) (Source: Ma, 2019)

Fig. 2.3 Chongqing People's Stadium (left); Tianjin People's Indoor Stadium (right) (Source: left, Yin, 1956; right, http://junbao123.blog.sohu.com/)

① Contemporaneously, early China-aided stadiums conveyed the socialist characteristics in their design languages, such as China-aided National Olympic Stadium of Mongolia, which will be demonstrated in later sections.

② This stadium was designed by Chongqing Architectural Design and Research Institute in 1953 and its construction was completed in 1956. As the first A-level stadium of PRC, its site area covers 120,000 m^2, and its gross floor areas are 96,000 m^2.

In fact, Chinese domestic stadium development continuously received influence from the Soviet Union. Most of the sports building standards adopted the Soviet standards in the early period of the establishment of PRC. In 1956, the Shanghai Industrial and Urban Architectural Design Institute translated the "Design Standard for Sports Structures (Outdoor Stadiums and Indoor Stadiums) H110-53" set by the "National Construction Committee of the Soviet Union's Secretary Meeting" published in Moscow. It set standards in site selection, layout, public facilities, and classified the stadiums based on capacity (Table 2.2), becoming the main standard basis for the design and construction of Chinese domestic stadiums. For instance, the scale of Chongqing People's Stadium referenced the "Soviet Standard of Large-scale Outdoor Stadiums", and set its capacity at 40,000 people (Yin, 1956).

Table 2.2 Categories of stadiums in "Design Standard for Sports Structures (Outdoor Stadiums and Indoor Stadiums) H110-53" ①

<table>
<tr><th colspan="5">Categories of Stadium Based on Capacity</th></tr>
<tr><td>Category</td><td colspan="2">Large Stadium</td><td>Medium Stadium</td><td>Small Stadium</td></tr>
<tr><td>Capacity (seats)</td><td colspan="2">20,000 to 50,000</td><td>5,000 to 20,000</td><td>Less than 5,000</td></tr>
<tr><th colspan="5">Categories of Indoor Stadium</th></tr>
<tr><th rowspan="2">Category</th><th colspan="3">Indoor Size (m)</th><th rowspan="2">Activity Capacity (number of people per class)</th></tr>
<tr><th>Length</th><th>Width</th><th>Height</th></tr>
<tr><td>Large Indoor Stadium</td><td>36</td><td>18</td><td>7</td><td>75</td></tr>
<tr><td>Medium Indoor Stadium</td><td>28</td><td>16</td><td>6.5</td><td>60</td></tr>
</table>

In the late 1950s, the unrealistic stadium development plans② emerged, when various provinces were required to construct large sport centers and stadiums. Beijing was even planning to build a 250,000-seat stadium at Wukesong, and Professor Ge Ruliang from Tongji University even proposed a design scheme for the stadium. Beijing Workers' Stadi-

① The table is based on the "Design Standard for Sports Structures (Outdoor Stadiums and Indoor Stadiums) H110-53", Beijing: Architectural Construction Press, 1956,16.

② In March 1958, China's State Sports Commission promulgated the program for the development of China's sports for ten years. China's ten-year plan for sports was formed, which required that "from 1958 to 1967, the number of public sports venues in the provincial cities would increase from 434 in 1957 to 1,023. The provincial cities would have more than one sports venue available for competition." (The State Sports Commission. Proposal on the ten-year planning of sports venues (draft), 2 October, 1958)

um (1959) is one of "Ten Great Buildings"① in Beijing at that time. This 62,000-seat stadium is China's first large stadium that holds over 50,000 seats, and also the largest modern stadium with the highest standard in China at that time (Fig. 2.4). The gross floor area is 87,080 m^2, 520 m from east to west and 693 m from north to south. The 69 m × 104 m central football field is surrounded by the 400 m-long track and oval stands (Zou, 2003). Ouyang Can from BIAD took the main responsibility of the architectural design, cooperating with other experts nationwide. Years later, Beijing Workers' Indoor Stadium (Fig. 2.5) was designed and constructed in the adjacent area, also designed by architects from BIAD. It was China's first large-scale indoor stadium with over 10,000 seats (13,500 seats), with a gross floor area of 42,000 m^2. During the period, the technicians "explored, designed and constructed at the same time"②; therefore, it only took one year to complete the stadium, from the project approval to construction accomplishment.

Beijing Workers' Stadium abandoned the pursuit of "national style" in form and balanced its identity of being national key projects with succinct modern language, with concise segmentation in the facades of a circular volume. Such an approach was inherited by the Beijing Workers' Indoor Stadium, using external columns to divide the cylindrical facade in a simple and regular way. Vertical lines became the emphasis, in correspondence with the outdoor stadium nearby. The ring beam was used for transverse segmentation to increase the sense of unity. The bottom section expanded to form the base, and the roof eave protruded horizontally to make the whole feeling simple and light. The vertical simplified facades formed the template for Chinese stadiums (e.g., the Capital Indoor Stadium, Beijing, 1968, Fig. 2.6) in major cities of the 1960s.

① "Ten Great Buildings" were proposed and constructed by the Chinese government for the 10th anniversary of the founding of PRC, including the Great Hall of the People, the Museum of Chinese History and the Museum of Chinese Revolution (two houses belong to the same building, now the National Museum of China), the Chinese People's Revolutionary Military Museum, the National Cultural Palace, the National Hotel, Diaoyutai State Guesthouse, the Prime Hotel (removed and rebuilt), the Beijing Railway Station, the National Agricultural Exhibition Center and the Beijing Workers' Stadium.

② The "rapid design and construction" was advocated at the time, and a model of construction encouraged by Chinese central government was "explored, designed and constructed at the same time". "Ten Great Buildings" became the successful examples, all of which were completed in one year with a total gross floor area of 673,000 m^2.

Fig. 2.4 Beijing Workers' Stadium (Source: Ma, 2019)

Fig. 2.5 Beijing Workers' Indoor Stadium

Fig. 2.6 The Capital Indoor Stadium, 1968 (Source: Ma, 2019)

2.2.2 Structural Development of Chinese Stadiums from the 1950s to 1970s

In the late 1950s, international exploration boomed for new architectural structures and new technologies. The requirement of rapid recovery and development of industrialization of China led to a great upsurge in the construction of industrial buildings. The designs of the industrial constructions became the main task, and the design capabilities of large span space laid the foundation for the structure development of Chinese stadiums later (Zou, 2003). In the 1950s, two-dimensional steel roof structures such as the truss arch steel frame and the three-hinged arch steel frame were combined with the traditional concrete roof structure and the concrete frame structure of the main body. Still, the span was limited to be generally below 50 m. In the 1960s, the space grid structure began to be explored and utilized in Chinese stadiums for expanding the span over 50 m. For instance, the Capital Indoor Stadium used the rectangular space truss structure (span: 99 m × 112.2 m), Nanjing Wutaishan Indoor Stadium (Fig.2.7) used the octagon space truss structure (span: 76.8 m × 88.68 m), and Shanghai Indoor Stadium (Fig.2.8) used the circular cantilever rack space truss structure (span of 110 m diameter). Shell space structures were also in trial development in this period. For example, the competition hall of Tianjin Indoor Stadium adopted a reticulated shell structure, Guangzhou Indoor Stadium (1957) combined a thin shell roof with reinforced concrete steel beams, and Beijing Tennis Stadium (1958) tried the hyperbolic flat shell structure. It is worth mentioning that suspension structures with relatively great technical difficulties were also in utilization. In the design of Beijing Workers' Indoor Stadium (1959), the double-layer suspension roof was first adopted in China, which enabled its total external diameter of the roof to reach 94 m. Innovations in the structure development of Chinese stadiums tended to be continually progressed as preparation for future stadium designs and constructions.

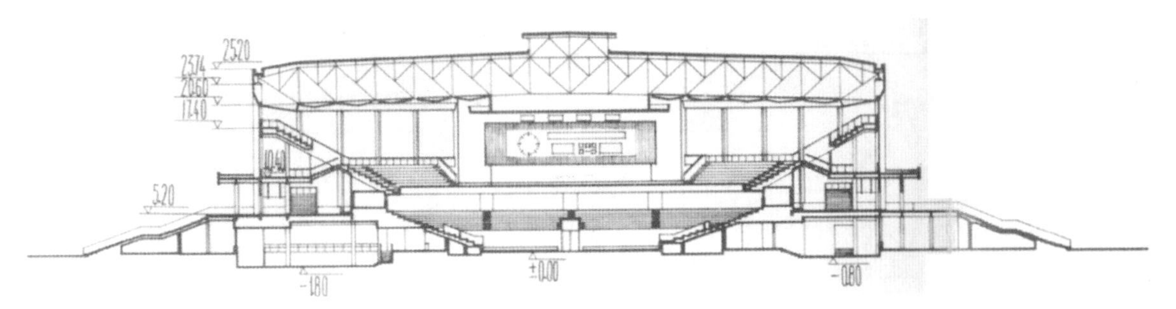

Fig. 2.7 Section of Nanjing Wutaishan Indoor Stadium (Source: *Architecture Illustrated — Nanking Wutaishan Gymnasium* (《建筑实录——南京五台山体育馆》), 1976)

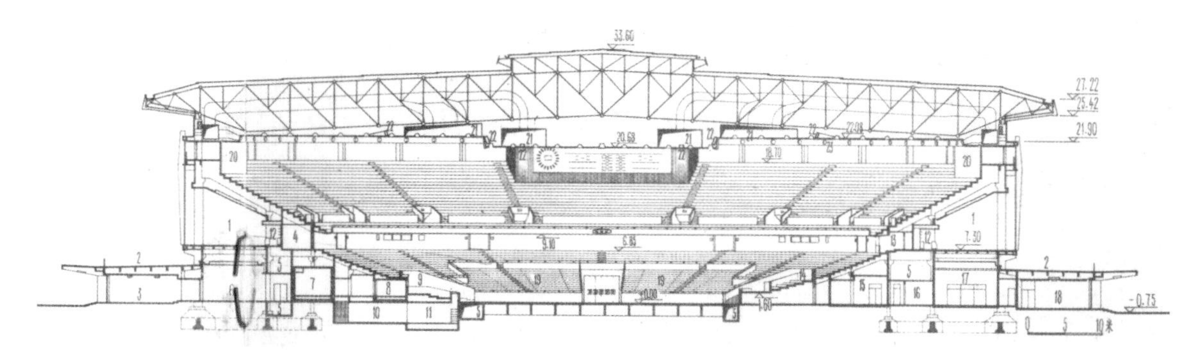

Fig. 2.8 Section of Shanghai Indoor Stadium (Source: *Architecture Illustrated — Shanghai Gymnasium* (《建筑实录——上海体育馆》), 1976)

2.2.3 Summary

Generally, for the first 30 years of PRC, Chinese domestic stadiums developed slowly (Table 2.3). The stadiums constructed in big cities of China became parts of the city's image and the witness of times. The features and development of Chinese domestic stadiums also affected China-aided stadiums overseas to a certain degree, as demonstrated in the following parts.

Table 2.3 Summary of the development of Chinese domestic stadiums from the 1950s to 1970s

Time	Scale (Seat)	Cases	Architectural Features	Structural Development
1950s	Indoor Stadium: 4,000–6,000	Beijing Indoor Stadium; Tianjin Indoor Stadium; Guangzhou Indoor Stadium	National forms; socialist realism	Two-dimensional steel roof structure combined with the traditional concrete roof structure and the concrete frame structure of the main body

continued

Time	Scale (Seat)	Cases	Architectural Features	Structural Development
1960s	Outdoor Stadium: 10,000	Beijing Workers' Stadium; the Capital Indoor Stadium;	Succinct modern language, with concise segmentation in the facades	Space grid structure (rectangular space truss structure, octagon space truss structure, etc.)
1970s	Indoor Stadium: 5, 000	Shanghai Indoor Stadium; Nanjing Wutaishan Indoor Stadium;	Larger scale; diversified layouts	Shell space structures, suspension structures (in the experiment)

2.3 China's Foreign-aided Stadiums from the 1950s to 1970s

As one significant part of China's sports aid in the developing areas, China-aided stadium projects were exported in the form of "complete projects", which means that, in "finance, design, construction and management", the Chinese government took full responsibility. However, probably due to the deficiencies of modern stadium (large span constructions) design abilities, Chinese stadium aid started from financial assistance, basic technique support and labour exportation, which was different from some other construction aids that Chinese architects were involved initially such as conference buildings, office buildings, schools, dwellings and hospitals. With the development of Chinese domestic stadiums, Chinese architects participated more in the designs of overseas aid stadiums. From the 1950s to 1970s①, 16 stadiums (Table 2.4) were assisted, designed or later constructed with the help of China. Although the number of the China-aided stadiums of the early period was relatively fewer and the scale was smaller, they expressed the architectural ideas of Chinese architects who spoke for certain positions of the history of Chinese sports architecture and modern Chinese architecture.

2.3.1 How the History Started

Asian countries received the earliest China-aided stadiums since the late 1950s, which

① The time of the stadiums in the study referred to the time when the design was completed because the study focuses on how the design developed under various periods. Sometimes the construction lasted for years or was shelved due to complex reasons, so it may be inappropriate to use the time of construction completion.

were exported as required by the recipient countries or from the donors' generous willing. The neighbouring country Nepal obtained the first China-aided stadium (Fig. 2.10) project in 1956, as one example of various kinds of financial and technical assistance to fulfil China's promise by providing aids to the developing world. Later, labour support was also requested by some countries for their lack of abilities to construct mega-structures, such as Mongolia for its National Olympic Stadium in Ulaanbaatar①, and Vietnam for its national stadium in Hanoi (Fig. 2.11).

Since the 1960s, more China-aided stadiums were exported for preparing sports events by the recipient countries, and the recipient regions expanded from socialist countries to nationalist countries. After the Second World War, some colonial and semi-colonial countries of Asia, Africa and Latin America became independent. China intended to support a new international sports event named the Games of the Newly Emerging Forces (GANEFO), including developing countries in Asia, Africa, Latin America and some other socialist countries② (Fan & Lu, 2012). In April 1963, one of China's leaders, Liu Shaoqi, visited Indonesia and signed with President Sukarno a joint declaration that China would support the GANEFO by providing Indonesia with an US $ 18 million gift for holding the first GANEFO (Fan & Xiong, 2005). Chinese government assisted the construction of a new national stadium in Jakarta with donated sports facilities and equipment (Fan & Lu, 2012; Fu, 2007).

However, the study found that, actually, the first GANEFO was held in the Gelora Bung Karno Stadium (GBK Stadium), the only stadium in Jakarta built in the 1960s and named after Indonesia's first president Suekarno (original name Koesno Sosrodihardjo, 1901-1970), who was an architect as well and whose design was known as the "Temugelang Design". The main concept of the stadium came from the President Sukarno to have a fully covered tribune which was distinct from the stadium in that era. The stadium was later designed by the famous Indonesian architect Frederich Silaban and initially funded through a special loan from the Soviet Union (Julius, 2004). Silaban was inspired by the large central stadium of Lenin, Moscow, and the Museum of Anthropology in Mexico. He wanted to build a stadium to protect all the spectators from the rain and heat, so the roof of the stadium was oval and continuous, contributing to a rhythmic display of unity and harmony ex-

① More about this stadium will be demonstrated in the following paragraphs of the study.

② 48 countries presented the 1st GANEFO in June 1963 including China, Cambodia, Guinea, Indonesia, Iraq, Mali, Pakistan, North Vietnam, the United Arab Republic (Egypt) and some other countries from Asia, Africa and Europe.

Table 2.4 Information on China's foreign-aided stadiums from the 1950s to 1970s

No.	Year	Continent	Country	Project Name	Aid Way	Design Institute	Capacity (Seat)	Photo
1	1956	Asia	Nepal	Dasarath Rangasala Stadium①	Financie and technology	—	25,000	
2	1958	Asia	Mongolia	National Olympic Stadium	Finance and construction②	BIADI	12,500	
3	1958	Asia	Vietnam	National Stadium in Hanoi	Finance and construction	—	10,000	
4	1963	Asia	Indonesia	Gelora Bung Karno Stadium (National Stadium)	Finance and construction	—	110,000③	
5	1966	Asia	Cambodia	Indoor Stadium (with an International Village) of the Olympic Sports Complex of Cambodia	Finance, technic support and construction	BIADI	8,000	
6	1968	Africa	Tanzania	Zanzibar Amaan Stadium	Finance, design and construction	BIADI	10,000	
7	1973	Africa	Ethiopia	Abebe Bikila Stadium	Finance, design and construction	BIADI	30,000	

① The Dashrath Rangasala Stadium is the only multi-purpose international stadium in Nepal built in 1956. Since its establishment, it has been the venue for all the major national as well as international events. The stadium has a capacity to house 25,000 spectators. The government of China provided financial as well as technical support to update lighting system, sound system, display screen and other categories of equipment of the stadium, so that it could meet the international requirements to host any events. The government of China spent more than 60 million just for the upgrading of stadium floodlights and for setting up a back-up generator. (https://xinhuanepal.tumblr.com/post/37395953916/xinhua-gallery-china-aid-project-dashrath)

② According to the official archive, the author inferred that Chinese technicians might participant in the designs or other technic supporting work for the China-aided National Olympic Stadium in Mongolia, but there was no direct evidence for that the mainly architectural design was accomplished by Chinese architects. Therefore, the "design" was not listed in the "Aid Way".

③ The capacity was reduced to around 80,000 in later twice renovations.

continued

No.	Year	Continent	Country	Project Name	Aid Way	Design Institute	Capacity (Seat)	Photo
8	1974	Africa	Uganda	National Stadium	Finance, design and construction	BIADI	40,000	
9	1975	Africa	Sierra Leone	Siaka Stevens Stadium	Finance, design and construction	Zhejiang Industrial Architectural Design Institute (ZIADI)	30,000	
10	1975	Asia	Syria	Damascus Tishreen Stadium (indoor)	Finance, design and construction	BIADI	7,141	
11	1976	Asia	Pakistan	Jinnah Stadium (Main Stadium of Pakistan Sports Complex)	Finance, design and construction	BIADI	50,000	
12	1976	Asia	Pakistan	Liaquat Stadium (Indoor Stadium of Pakistan Sports Complex)	Finance, design and construction	BIADI	10,000	
13	1977	Africa	Somalia	Mogadishu Stadium	Finance, design and construction	Guangxi Comprehensive Architectural Design Institute①	30,000	
14	1977	Africa	Benin	Friendship Stadium (Cotonou Sports Comple)	Finance, design and construction	Shanghai Industrial Architectural Design Institute (SIADI)	30,000	
15	1977	Africa	Benin	Friendship Indoor Stadium (Cotonou Sports Comple)	Finance, design and construction	SIADI	5,000	

① This project was assigned by China's Sports Commission to Guangxi Province, so it was designed by Guangxi Comprehensive Architectural Design Institute (Guo, 1983).

continued

No.	Year	Continent	Country	Project Name	Aid Way	Design Institute	Capacity (Seat)	Photo
16	1977	Africa	Senegal	Stade de l'Amitié (Stadium of Friendship) (Renamed) Stade Léopold Senghor	Finance, design and construction	Guangdong City Architectural Design Institute	60,000	

Fig. 2.10 Dasarath Rangasala Stadium in Nepal (Source: http://stadiumdb.com/stadiums/nep/dasarath_rangasala_stadium)

pressed in the stadium. China helped with the upgrading constructed by the labourers exported from China. This 110,000-seat stadium demonstrated well the socialist exportation through stadium aids from the two largest socialist countries of the era. (Fig. 2.12, Fig. 2.13)

In 1966, China helped Cambodia to build an indoor stadium① with 50,000 seats as well as other facilities for holding the 1st Asian GANEFO. The stadiums in Pakistan Sports Complex② were also designed and constructed by China for Pakistan's holding of the 8th Asian Games. Most of the China-aided stadiums, generally, were closely associated with major regional or international sports events. These stadium aids from China were also parts

① This stadium will be illustrated in the following sections.

② This stadium will be illustrated in the following sections.

Fig. 2.11 Ho Chi Minh in the newly built Vietnam National Stadium (top left); a Vietnamese stamp with the newly built Vietnam National Stadium as the background (top right); the National Stadium of Vietnam in a Vietnamese newspaper in 1958 (bottom) (Source: top left and bottom, from: https://m.vff.org.vn/ben-le-san-co-746/75-nam-san-septo--hang-day--ha-noi-dai-hoa-dep-giua-thu-do-12525.html?fbclid=IwAR1uCHyjC3dcya8CgNEDSY193n6FD-PRwualz1ngta1oxYUS1WFCexLs-y-A; top middle, from CADRG; top right, from the author's collection)

Fig. 2.12 Title page and other pages of the *Journal of China Pictorial*, special volume for the GANEFO in 1963

Fig. 2.13 GBK stadium under construction (left); GBK stadium after construction (right) (Source: https://oumagz.com/ou-chill/gelora-bung-karno-dari-dulu-hingga-kini/)

of its sports aids, which were initially used to support the national liberation movements in developing countries in Asia and Africa (Yu & Yuan, 2010). As one significant part of such diplomacy, the journey of designing and constructing stadiums overseas started thereby.

2.3.2 Pilot Designs after Domestic Stadium Practice

Based on the database① (Appendix 2) of the study, it was not until the late 1960s that China's stadium aid transferred from simply financial and labour assistance to the actual designing involvement of Chinese architects, such as Zanzibar Amaan Stadium (1968) in Tanzania and Abebe Bikila Stadium (1973) in Ethiopia. From the 1960s to 1970s, seven outdoor stadiums and three indoor stadiums were designed by Chinese state-owned design institutes. Most of the institutes were affiliated to the Chinese Ministry of Construction② located in Beijing and other major cities of China, as arms of large-scale socialist construction. Chief architects of these institutes took the general responsibilities for the design work.

The scale of China-aided stadiums overseas (outdoor stadiums, around 30,000 to 50,000 seats; indoor stadiums, no more than 10,000 seats) were relatively smaller than those in major Chinese cities. Another differentiation was that the number of China-aided outdoor stadiums was more than that of Chinese domestic ones, the latter of which halted after the construction of Beijing Workers' Stadium in 1959. And the structures of China-aided outdoor stadiums of this period inherited the concrete frame structure used in Beijing Workers' Stadium. Still, most of them owned smaller areas of stand-coverage roof③. Most of the China-aided outdoor stadiums had two floors under one single platform space, while the fa-

① The database was collected and concluded based on resources from the Chinese official websites and publications, the achieves of the enterprises involved, and other reliable Internet websites.

② BIADI, ZIADI and SIADI were all affiliated to the Chinese Ministry of Construction.

③ The Beijing Workers' Stadium possessed full coverage roof of the whole platform. For China's foreign-aided stadiums of this period, only the Uganda National Stadium had the overall coverage roof.

mous Beijing Workers' Stadium had four floors with two-layer stands (Fig. 2.14). Most China-aided indoor stadiums used two-dimensional steel frame roof structures, while Chinese domestic indoor stadiums tried more complicated structures such as diversified forms of shells. For this period, the overseas aid stadiums could be regarded as the lower standard version exportation of Chinese domestic stadiums.

Fig. 2.14 Beijing Workers' Stadium, 1959 (left); Siaka Stevens Stadium, Sierra Leone, 1973 (right) (Source: left, from Ma, 2019; right, from https://cocorioko.net/welcome-to-the-siaka-stevens-stadium/)

If comparing China-aided stadiums with Chinese domestic ones in architectural perspectives, similarities and dissimilarities coexisted simultaneously. Although located in different areas and countries, the basic functional requirements of the designs were generally the same as those of Chinese domestic stadiums, which were normally made by Chinese government and architects since most recipient countries were in low developed level and their understanding of modern sports architecture was insufficient. In addition, the facade stylish shared the common sense of frugal modern expression developed from the socialist realism with concise segmentation by structure elements. Both China-aided outdoor stadiums and the Chinese domestic ones shared parallel vertical segmentation by supporting concrete frame pillars except for that the Beijing Workers' Stadium covered the outsider with walls and windows for rooms under the stand space, while China-aided stadiums exploded the two layers of pillars without utilization of the under-platform space. All three overseas indoor stadiums were designed to be basic cube shapes with heavy roofs (generated by structures) and vertical segmented facades. However, the homogeneity of indoor stadiums was decreased due to the diversified handling of walls, windows, patterns and other details, as initial regionalist design approaches of China-aided stadiums (Fig. 2.15, Fig. 2 .16).

Beijing Workers' Stadium

VS.

Jinnah Stadium in Pakistan

National Stadium of Uganda

Fig. 2.15 Comparison of the Chinese domestic stadium and China's foreign-aided stadiums (outdoor) (Source: drawn by the author)

Beijing Workers' Indoor Stadium

The Capital Indoor Stadium

VS.

Liaquat Stadium in Pakistan

Damascus Indoor Stadium in Syria

Friendship Indoor Stadium in Benin

Fig. 2.16 Comparison of Chinese domestic stadiums and China's foreign-aided stadiums (indoor) (Source: drawn by the author)

The similarities and dissimilarities attract the author's attention and appear to be an integral aspect of the history of Chinese sports architecture and modern Chinese architecture. How the Chinese architects designed under the complex circumstance, and special foreign aid policies, and what Chinese architects were mainly concerned about in these architectural exportations are the main focuses of this study and will be explicated in detail in the following subsections.

2.3.4 Case Studies

For this period, three stadium cases are chosen for the case studies, National Olympic Stadium of Mongolia, National Sports Complex of Cambodia and the stadiums of China-aided Sports Complex in Pakistan. The first two stadiums are representative of the early stadium aids from China with labour and technical support exportations. The following case, China-aided sports complex in Pakistan, is the only sports complex that China exported during this period, and it represents the highest architectural design level among early China's foreign-aided stadiums. These cases illustrate the transforms from labour/technical, aids to design aids in China's stadium exportations.

2.3.4.1 Labour and Technique Support Exportation: National Olympic Stadium of Mongolia and National Sports Complex of Cambodia

In early times, there was little involvement of Chinese architects in the design process, as the stadiums were designed by the local architects. In this situation, China mainly served as an assistant with exported financial, technical and labour supports. In this section, two early China's foreign-aided stadiums are studied to illustrate Chinese aid of the early period, about how the aid played roles and how the aid influenced future designs of the stadiums.

National Olympic Stadium of Mongolia

National Olympic Stadium of Mongolia was the first stadium that China provided labour support abroad, which formed the base of labour exportation of China's foreign aid. The stadium was constructed when China and Mongolia kept close diplomatic relationship, when top heads of the Chinese government visited Mongolia frequently, such as the Premier Zhou's two visits in 1954 and 1960[①]. In fact, the labor assistance was initially conducted to satisfy the request from the Mongolia government, which was made in 1950 soon after the

① China's Premier Zhou Enlai paid visits to Mongolia in 1954 and 1960. The vice president Zhu De also visited there in 1956.

establishment of PRC, for exchange of overseas Chinese returning to China[①]. Five years later, Xi Zhongxun[②] signed the agreement telegram (Beijing City Archives, 1955) that opened the way of labour assistance to Mongolia for the exchange. In 1956, more financial, technic and labor aids were agreed for the construction of factories, infrastructures, and civil buildings in Ulaanbaatar with the total value of 160 million roubles (approx. RMB 374 million yuan) and thousands of technicians and labourers' exportation from 1956 to 1959 (Beijing City Archives, 1956). The construction was under the charge of Beijing Construction Engineering Bureau, the design work was taken on by Shanghai, and the labourers were sent from Hebei Province of China (Hebei Province Archives, 1957; Gu, 2015). These assistances helped the recovery and construction of the city Ulaanbaatar, among which the National Olympic Stadium was financed and constructed with China's aid in 1958 for the improvement of Mongolia's sports infrastructure and athletes' training conditions (Yu & Li, 2016) (Fig. 2.17).

Fig. 2.17 The newly constructed National Olympic Stadium in Ulaanbaatar, 1958 (Source: Mongolian Study Association of Chinese History and Culture, 2019)

Designed by Soviet architects, this 12,500-seat stadium used concrete frame structure in the form of circular walls. In addition, there was only one layer of the stand, half of

① Due to the Second World War, a number of Chinese were retained in Mongolia and their request of returning the motherland was rejected firstly by the Mongolian government over concerns of a labor shortage. (Archives of China's Ministry of Foreign Affairs, 106-00025-03)

② Xi Zhongxun, secretary-general of the state council of China during the time.

which got covered by the plane steel structure roof. The cultural and regional elements were expressed frankly in the shapes and patterns of the windows and doors, and the colours of the walls, the cornice and the roof. The main entrance hall absorbed the popular nationalism from Soviet Union's architecture, and the colours, detail and format of the "palace" adopted Mongolian landmark buildings (Fig. 2.18) that were transformed from the dominant Russian period. It coincidentally shared similarities with Chinese domestic stadiums of the same period, under the influence of the Soviet Union.

China exported numbers of workers to complete the construction. These workers were selected carefully with high skills, many of whom even held the "Ninth Grade" qualifications of construction skills[①], the highest standard in China of that period. With good quality of China's labour help, this stadium was constructed well and kept continually functioning for ages. The original structure and design might have limited the upgrading and renovation of this stadium since a second floor was added above only parts of the platform later for expanding of functions and equipment. However, as the only large stadium in the city centre of the capital, the significant national events and sports games were instantly held here, such as the biggest Mongolian traditional festival "Naadam" . The modernization of the surrounding area was also continuing (Fig. 2.19). Another football stadium was constructed nearby for private football team training, together with modernized office buildings and dwellings (Fig. 2.20). The stadium becomes one activating point of regional development and plays a role in the process of urban modernization of Ulaanbaatar.

① This fact was stated by Mr. You Baoxian in the interview, the former chief director of China Architecture Design Institute, who once participated in numbers of China-aided projects.

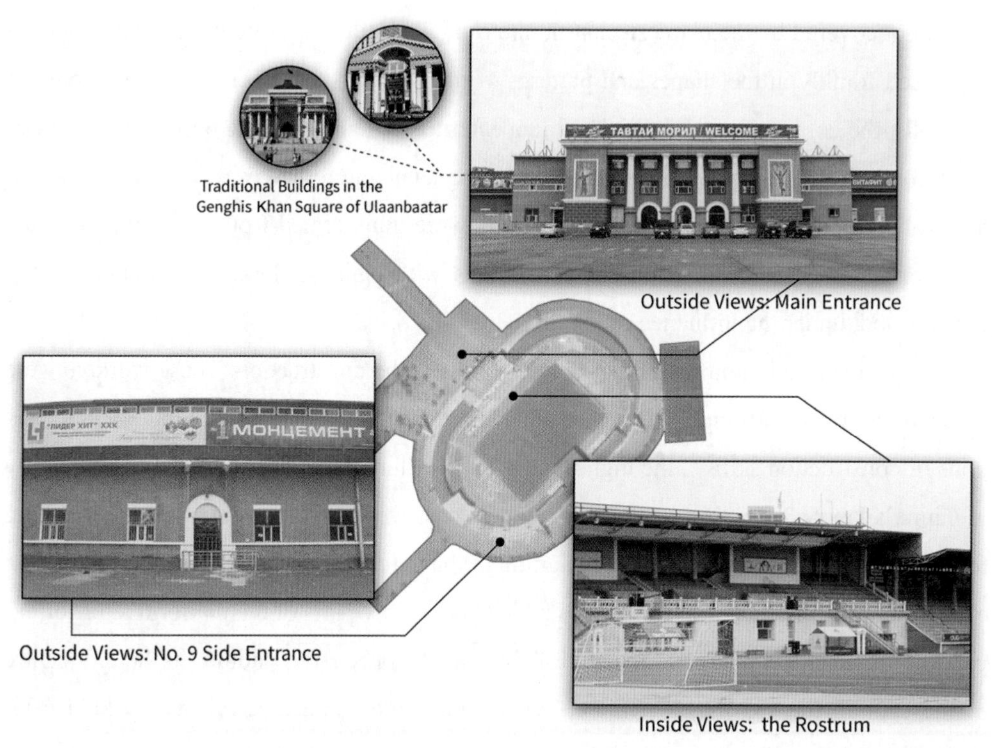

Fig. 2.18 Outside and inside views of National Olympic Stadium in Ulaanbaatar, 2019 (Source: the photos were taken by the author in 2019)

National Olympic Stadium of Cambodia

Cambodia was one of the first batch of countries in Asia that received China's economic and technical assistance. Faced with the Cold War confrontation of two camps, more and more emerging Asian and African nationalist countries, such as Cambodia, began pursuing a foreign policy of peace, neutrality and non-alignment for the sake of maintaining national independence and state sovereignty. To achieve economic independence and development, Cambodia began to formulate two-year economic development plans (1956-1957), and decided that rather than rely solely on U. S. and France, it openly expressed willingness to accept assistance from any country. As an immediate respond, China expressed the willing to provide economic and technical assistance to Cambodia without attaching any conditions or privileges (CCCPC Party Literature Research Office, 1997). China offered free supplies and technical assistance to Cambodia in 1956 and 1957 to support its first "Five-year Plan", with a total amount around 800 million riels (equivalent to 8 million pounds, or about 55.14 million RMB) (Ministry of Foreign Affairs of China, 1958). Besides, the economic aid increased after Cambodia's announcement of breaking diplomatic relations with

Fig. 2.19 "Naadam" event held in the National Olympic Stadium of Mongolia, Ulaanbaatar (Source: https://www.amazingmongolia.com/the-festival-in-mongolia-2018/)

Fig. 2.20 The surroundings of the National Olympic Stadium of Mongolia (Source: drawn by the author with photos taken by the author in 2019)

U.S.[①] With the support from China, the National United Front of Cambodia led by Norodum Sihanouk (1922-2012)[②], the National Unity Government of Cambodia with Penn Nouth as the Prime Minister, and Khieu Samphan as Deputy Prime Minister were established in Beijing. Sihanouk pointed out that "China gave us a lot of generous, unconditional brotherly support that enabled us to achieve a historic victory" (Wang, 1999). During Sihanouk's reign, the construction of the Phnom Penh was implemented with ambitions to quickly restore the city and establish a new capital with large-scale modernization and urbanization. These constructions were designed by Cambodian architects[③] who returned from abroad and were strongly supported by China. From 1964 to 1970, China exported assistance to help Cambodia with the construction of three factories and one national stadium, one international village for sports athletes, laboratories for Kampong Cham Royal University (12 labs and an internship workshop, built in 1968), a hospital building (with 200 beds, built in 1969) and other complete projects (Shi, 1989). According to Chinese statistics, from 1956 to 1969, China provided Cambodia with economic assistance worth more than RMB 200 million yuan and military assistance worth RMB 36 million yuan (Wang, 1999). China's aid helped to shape this emerging city, and some projects became representatives of Cambodia's New Khmer Architecture.

To help Cambodia host the 1st Asian GANEFO[④], Chinese government provided aid, including materials, technical and labour supports, for the construction of an International Village and an indoor stadium covering an area of about 19,000 m^2 (completed in October 1966). Actually, Cambodia decided to construct its national Olympic Complex early in the 1960s, and was supposed to design, finance and construct it all by itself. The complex was

---

① U.S. continuously put pressure on Cambodia by ceasing to provide aid on the one hand and instigating South Vietnam and Thailand to impose an economic blockade on Cambodia, triggering border incidents, violating Cambodia's territory and airspace and bribing the domestic separatists and opposition forces in Cambodia, as well as conducting interference and subversion. In November 1963, Cambodia announced that it refused all assistance from U.S., and announced in May 1965 that it was breaking off diplomatic relations with U.S.

② Cambodian National Liberation Army, with its growing strength, liberated Phnom Penh in April 1975, overthrew the Lon Nol régime, and won the final victory of the war against U.S. aggression and for national salvation.

③ Most projects were designed by Vann Molyvann (1926-2017), an architect who had returned to Cambodia from France.

④ The second GANEFO was scheduled to be held in Cairo, and the second choice was Beijing. In 1964, the United Arab Republic asked China to help to build the stadium, or it would give up the GANEFO. The agreement was failed to reach for the high amount of money. Beijing decided to undertake the mission and started the construction of stadiums, including the capital stadium. Indonesia was hostile to China after its coup in 1965. Facing the opposition from the IOC, the name of the games was changed to be the 1st Asian GANEFO and was finally held in Cambodia.

designed and under construction initially for applying to host the Southeast Asian Peninsular Games in 1963, but the games were eventually awarded to Indonesia instead. Years later, the projects experienced economic and construction difficulties that delayed the completion time, after which Chinese government sent its assistance to solve the problems.

Fig. 2.21 Stamps of China in memory of the 1st Asian GANEFO, 1966. (Source: the photos were from the author's collection)

National Olympic Stadium of Cambodia was also designed by their famous local architect, Vann Molyvann (1926-2017), who studied at École des Beaux-Arts in Paris, before becoming a follower of Modernist architects Le Corbusier and Paul Rudolph. He returned to Cambodia and was appointed by King Sihanouk as the State Architect and the Head of Public Works in 1957 after the country was awarded independence from France. During the reign of King Sihanouk, from 1954 to 1971, Molyvann designed a series of landmarks in Phnom Penh, including the Independence Monument, Chaktomuk Conference Hall, White Building and Teacher Training College, etc. His works were among the most important collections of postcolonial buildings in Cambodia, referred to as the "New Khmer Architecture", the period of which is now referred to as Cambodia's Golden Age. Influenced by Le Corbusier's simplicity and modernity, Molyvann chose to combine the indoor stadium and the outdoor venue directly closed together, and located the outdoor stadium, indoor stadium and swimming pool in symmetrical layout. The long axis was in the west-east direction with the main entrance (also the entrance of the indoor stadium) of the site facing the west, for consideration of the predominant wind direction of the local in hot weather to achieve better air flow (Fig. 2.22).

The 50,000-seat indoor stadium had a square layout and identical facades for the four sides, where the vertical repetition of architectural features existed as commonly used in the architect's works from large-scale public buildings to smaller, anonymously designed private houses. Khmer decorative elements were used in the entrance facades and the roof of the rostrum area in the context of post-colonial Cambodia under Norodom Sihanouk's leadership (Fig. 2.23). Internal spaces within the stadium were divided into smaller-scaled zones for functions and structures.

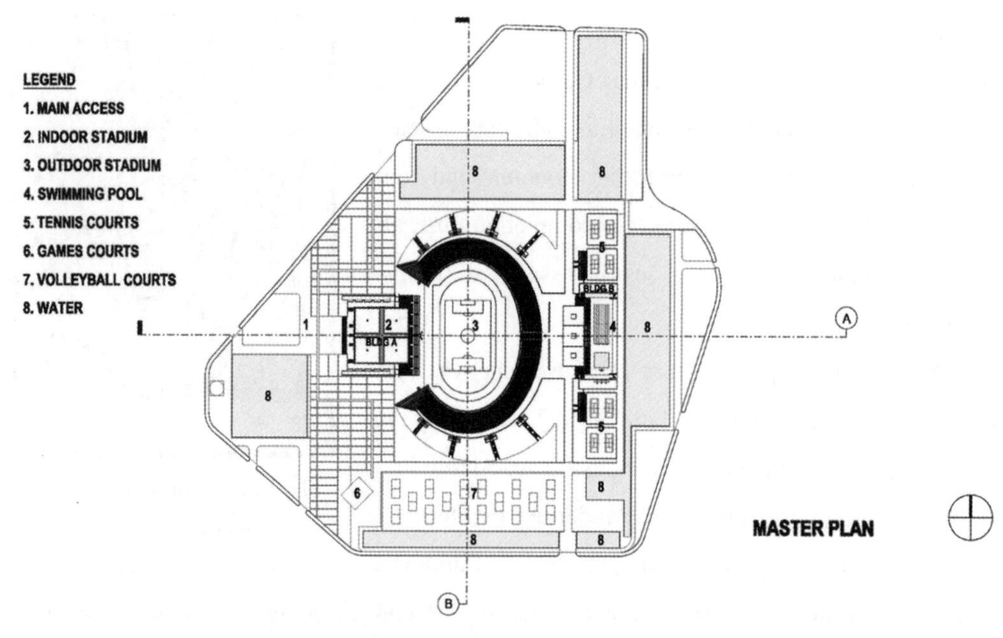

Fig. 2.22 Site layout of the National Olympic Complex of Cambodia (Source: http://www.vannmolyvannproject.org/the-national-sports-complex/)

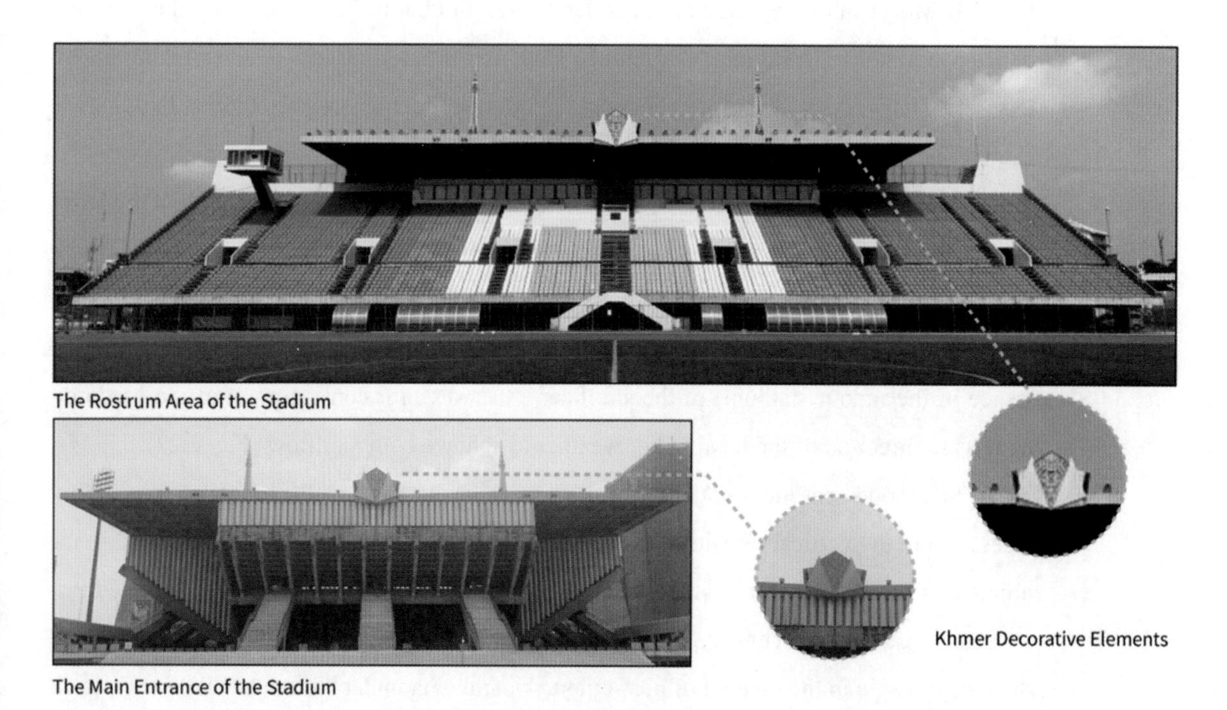

Fig. 2.23 Stadiums of National Olympic Complex, Phnom Penh, Cambodia (Source: the photos were taken by the author in 2019)

The roof of the indoor stadium was covered by four two-way slabs, which were then supported by 1,500 mm thick waffle beams to reach 40 m span, by using the reinforced concrete structure as most stadiums did during the time. Only four columns were put inside to support the complex roof concrete structure. The one-layer stand of the outside venue was exposed except for the rostrum areas, covered by cantilever structure from the back façade of the indoor stadium. The structural design was made by engineers from Corbusier's group but left serious difficulties for Cambodia to complete the construction, especially the indoor stadium's four-column supporting structure. The Chinese government sent a group of 30 technicians from BIADI to design and instruct the construction on site. Special lifting-up techniques were utilized to successfully complete the mission[①]. With construction help by labours from Chinese construction companies in Dalian, the complex was finally opened in 1964 and was in good quality and constant use ever since (Zhu & Wu, 2018) (Fig. 2.24, Fig. 2.25). The four-column supporting structure of the indoor stadium and the lifting-up technique also inspired Chinese architects in their future designs of China-aided stadiums[②] and Chinese domestic stadiums[③]. The technical input brought into China's domestic stadium practices from the experiments in the foreign-aided stadiums was unique and interesting.

Fig. 2.24 The National Olympic Indoor Stadium of Cambodia under construction, 1964 (Source: https://journals.openedition.org/abe/3615)

① From the interview with You Baoxian, who was involved in the aid work of the stadium.

② For instance, the indoor stadium of China-aided Pakistan sports complex.

③ For instance, the Capital Indoor Stadium and Shenzhen Indoor Stadium.

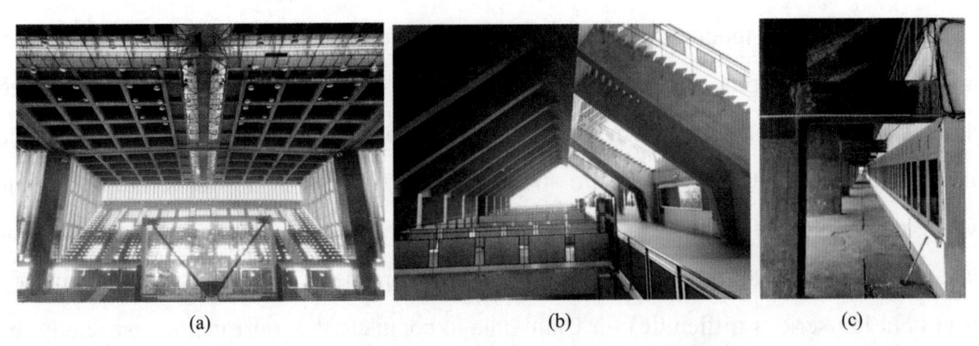

(a) (b) (c)

Fig. 2.25 Structures in National Olympic Indoor Stadium of Cambodia: (a) interior view of the indoor stadium; (b) outside view of the structure of the connection between indoor stadium and outside stadium; (c) support structure of the platforms of the outside stadium. (Source: the photos were taken by the author in 2019)

The natural ventilation technologies utilized in the stadium were worth mentioning. The vents under the seating area of the indoor stadium provided natural ventilation from the three sides, and daylight and heat remained at a fairly low level with the use of vertical interlocked angled sheets of metal that did not hinder ventilation. Such design naturally formed the rhythmic patterns of the facades, contributing to the special image of the stadium (Fig. 2.26). The attention to coeval architectural forms and broader regional intersections from the architect impressed Chinese architects① and such approaches and considerations were later utilized in both Chinese domestic stadiums② and China's foreign-aided stadiums.

The National Sports Complex in Phnom Penh became one of the city's largest structures, and one of its busiest and most beloved public spaces (Nelson, 2017). It was listed as one of the world's irreplaceable treasures in 2016 by the World Monuments Fund. In their official statement, the organization called the stadium "an iconic symbol of the massive post-independence effort that transformed Cambodia from an agrarian colony into a modern state", and suggested that "the design is one of the most important examples of regionally inflected modernism"③. Although the stadium was not designed by a Chinese architect, it could not have been completed without the technique and labour aids from China. The economical and elegant solutions remained appropriate for stadiums in the hot climate. In addition, its attributes of modernist tendencies with regional consideration were shared by many Chi-

① As stated by You Baoxian and Zhou Qinglin in the interviews, the design of National Olympic Indoor Stadium inspired their design of China-aided Stadiums in Pakistan.

② For instance, Guangxi Indoor Stadium used a similar design approach for natural ventilation.

③ https://www.wmf.org/project/national-sports-complex-cambodia.

Fig. 2.26 Vents under the seating area of National Olympic Indoor Stadium of Cambodia (Source: the photos were taken by the author in 2019)

na-aided stadiums later (Fig. 2.27). Chinese architects might have been influenced consciously or unconsciously in the process of participating and learning, and become better prepared for future designs. The author regards this case as the pre-declaration of China-aided stadiums.

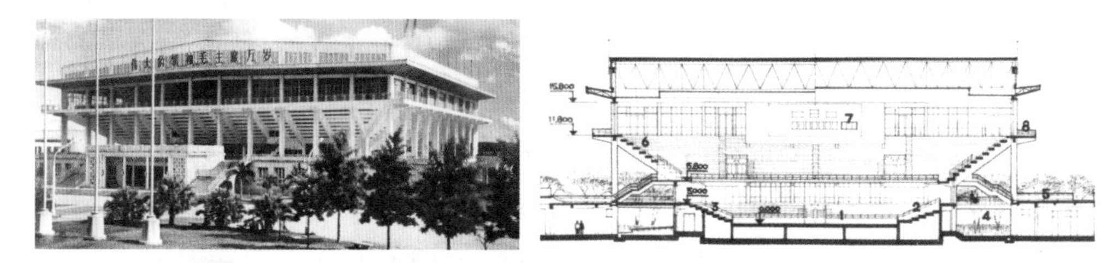

Fig. 2.27 Guangxi Indoor Stadium (Source: Architectural Illustrated — Guangxi Gymnasium in Nanning《建筑实录——广西体育馆》, 1976)

2.3.4.2 Design with the Sense of Mission and Honor — China-aided Sports Complex in Pakistan

Pakistan started a diplomatic relationship with China since 1950 and was one of the first batch of countries to recognize PRC(Han, 2014). China and Pakistan have developed into a permanent partnership ever since, and Pakistan has always firmly supported China on significant international issues. In 1974, Islamabad, the capital of Pakistan, won the opportunity to host the 8th Asian games (1978). However, at the time Islamabad was a newly built capital with insufficient sports facilities, and the Pakistan government decided to construct a sports complex to hold sports events such as the Asian Games. China agreed to assist in the design and construction of the sports complex as well as giving financial supports (You, 2017).

Pakistan Sports Complex covers an area of 586,815 m^2. It includes a 50,000-seat outdoor stadium, a 10,000-seat multi-purpose indoor stadium and other facilities (swimming pool, training gym and athletes' hotel), with a gross floor area of 71,260 m^2. It was the first time, and may be the only time, that the site planning and architectural design of one sport complex were assigned to one design institute①, the BIADI. This helped to unify the planning and architectural styles in economical and practical ways. One of the leaders of the institute, You Baoxian was the general project director of the sports complex, together with other experience architects and engineers to fulfil the job, including architect Wang Tianxi, who produced the schematic design.

Located on flat green land in the suburban region of the capital, where the expressway expanded to the city centre, the site of the sports complex was arranged mainly with consideration of transportation. The main entrance was set on the north side of the site with the main stadium, the long axis of which conformed to the north-south layout requirements as most stadiums commonly did (Fig. 2.28, Fig. 2.29). The indoor stadium is located in the northeast side of the main stadium, the swimming pool southeast, and the athletes' hotel northwest. The general layout of the sports complex did not form an obvious axial picture and was relatively unconstrained. This feature can also be seen in the other China-aided sports complexes, such as the Benin sports complex. It was also consistent with the usual layout of Chinese domestic sports center of the early period. In later Chinese domestic sports complexes' layouts, the axial trends became more obvious, under this influence the layout of China-aided sports complexes changed simultaneously (see Chapter 3).

The outdoor stadium② covered a gross floor area of 41,600 m^2, with a platform of the four-point centred oval shape, a football field and a 400-meter track around, similarly with

① Normally for Chinese sports complex project, the general site planning was in charge of one design institute while the architectural designs of individual stadiums were in other institutes' purviews.

② Later the stadium was named after Muhammad Ali Jinnah, to be the Jinnah Stadium. Muhammad Ali Jinnah was a national hero of Pakistan, who contributed a lot for Pakistan's independence and was honored as the "father of Pakistan".

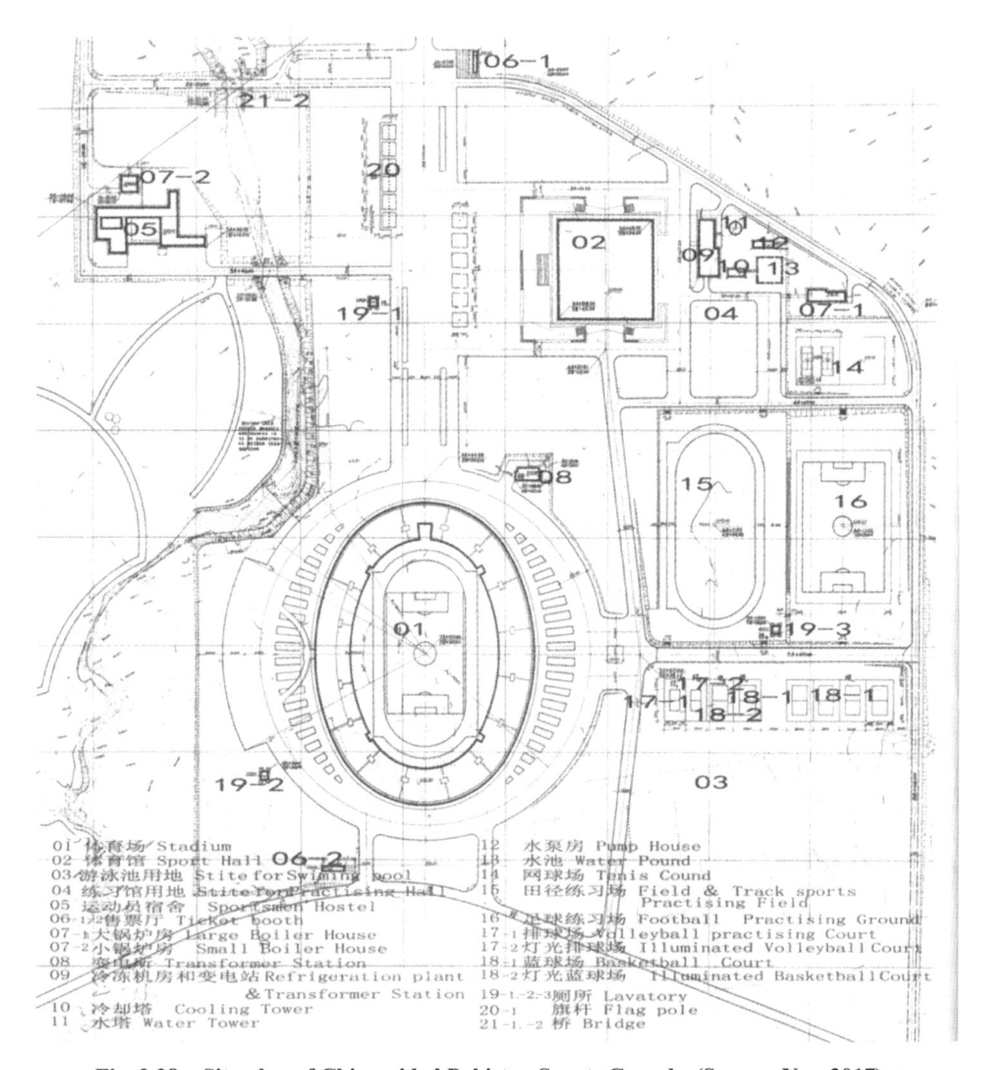

Fig. 2.28 Site plan of China-aided Pakistan Sports Complex(Source: You, 2017)

Fig. 2.29 Design drawing of Pakistan Sports Complex. (Source: Pakistan Sports Complex, poster color rendering by Wang Tianxi, 1976. From *Architectural Drawings*, 1979)

the Beijing Workers' Stadium. The triangular space under the platform was used for assistant rooms, activity and communication spaces, while outsider was suspended for adapting to the local climate, producing shadows and natural ventilation. The geometric decoration above the entrances of the stadium was intended to express local characteristics, but the effect was not obvious. The reinforced concrete frame structure was directly exposed, and so was the one-layer stand except for the rostrum part covered by a steel structure cantilever roof. Overall, the stadium illustrated Chinese architects' preference of the pure expression of modest modernism with a pragmatic consideration of the regional climate. (Fig. 2.30)

The indoor stadium① had a gross floor area of 19,685 m^2. Its square floor plan was constituted with a compact layout of functional spaces around the central playground, which can hold various sports competitions such as in basketball, volleyball, gymnastics and other competitions. The facade design took the local climate into account: the large walls occupied large areas to prevent direct sunlight into the indoor competition area; the four corners that did not affect the light-environment used glass windows; and the two forms of classical modernized architectural language of entity and emptiness were employed. The thick roof covered a span over 90 m of the steel grid structure, which was a breakthrough compared to Chinese domestic stadiums of the period (Fig. 2.31). To solve the

① The indoor stadium was later named after Liaquat Ali Khan, the first premier of Pakistan, to be the Liaquat Stadium.

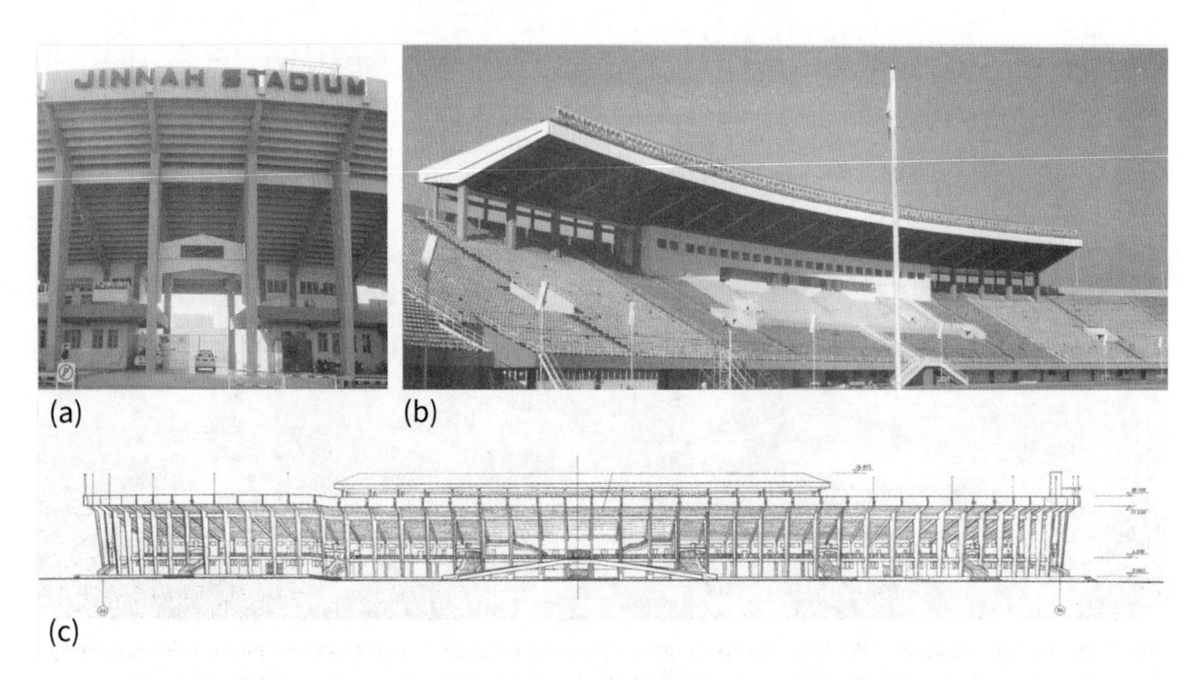

Fig. 2.30 Pakistan Sports Complex: (a) entrance; (b) inside view; (c) facade (Source: You, 2017)

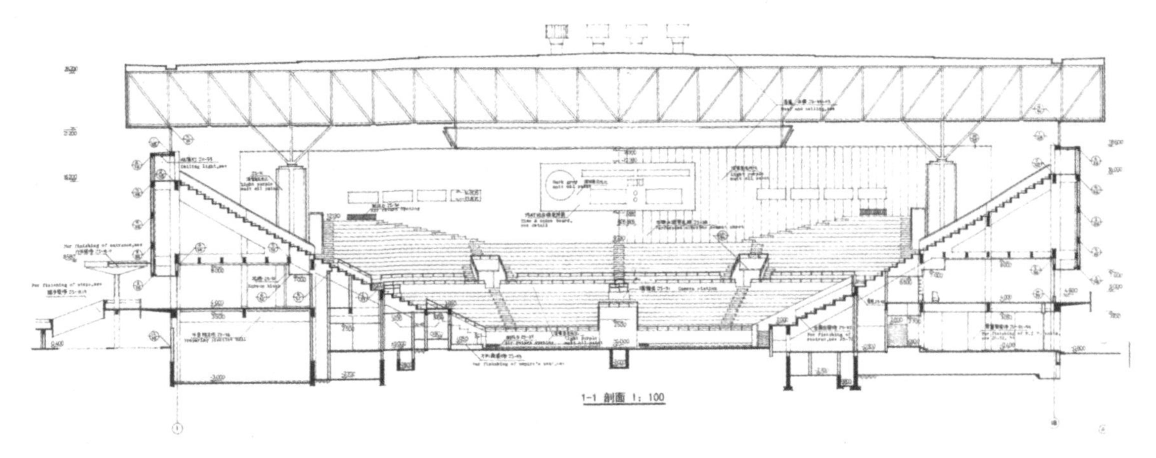

Fig. 2.31 Indoor stadium of China-aided Pakistan Sports Complex (facade and section) (Source: You, 2017)

structural problem, Chinese technicians created the four-column support grid system inspired by the indoor stadium of Cambodia's Olympic Complex that they had once been involved in. Such structure had never been used in Chinese domestic stadiums, in which normally the roof was supported by surrounding facade walls. The jack-up technology was used for the first time (Fig. 2.32). The successful application of the four-column supported grid structure and the construction technique was later applied to Shenzhen Indoor Stadium and the China-aided indoor stadium in Barbados (1992). The early attempts of China's foreign-aided stadiums sometimes promoted the development of Chinese domestic stadiums.

The designs of the sports complex were completed in 1976. Still, due to the economic situation of Pakistan, the 8th Asian Games was not held in Islamabad[①] as scheduled, so the construction was postponed. It was finally completed in 1984 (the swimming field was

① The 8th Asian Games was finally held in Bangkok, Thailand from 9 December to 20 December in 1978.

Fig. 2.32 Photos of on-site construction the indoor stadium of China-aided Pakistan Sports Complex. (the jacking construction techniques) (Source: You, 2017)

completed in 1988) and has been an important venue in Islamabad since. For instance, it hosted two South Asian Games, in 1989 and 2004, as well as the torch relay for the 2008 Beijing Olympic Games. It was a representative of China' stadiums in the early stage. It showed the working mode and designing level of Chinese architects, and their willingness of pursuing free creation under the modernity ideology. It became one of the paradigms for Chinese sports complex, as listed in China's most widely spread *Architectural Design Handbook* (Anonymous, 1994).

2.4 Summary: What Truly Mattered in the Designs

The 1950s-1970s period was the first historic stage of PRC, and also the first stage of the development of China-aided stadiums overseas. The stadium aid transformed from financial, technique and labour exportation of the early time to design exportation of Chinese architects later on. Although these stadiums were in relatively smaller scales and lower standards compared with Chinese domestic stadiums, at the same time, they owned their unique attributes and reflected Chinese architects' design tempts under complex circumstances, as indispensable supplements of the histories of Chinese sports architecture and modern Chinese architecture.

In this period, Chinese foreign aids policies mainly affected the countries/regions distributions that received China's stadium aids. Because the donor attached great importance to foreign aid as diplomatic tools for its international standing, the design institutes and designers who undertook the design tasks had a heavy sense of mission and national loyalty. The assignment to state-owned design institutes under the planned economy also made the personnel and organizations involved in China's foreign-aided stadium projects relatively fixed, which generated certain impacts on the architectural design. The mechanism of design and management was quite direct and simple. MERFC and the Ministry of Construction set up the location, the scale and the category, but the whole design content and processes were in the hands of design institutes and Chinese architects. The policy and mechanism of foreign aids at this stage did not have much influence on them.

Both architectural technique and function were the main focuses for Chinese architects when designing the overseas stadiums. To hold significant sports events was normally the purposes of exporting these mega-structures. Therefore, to meet the requirements of holding international games was put in the first place. Since the development of Chinese domestic stadiums was in the early period, and the architectural technique such as the structures did not experience enough improvement. New forms of structure were only in the exploration stage in Chinese domestic stadiums, and few had been put into practice. The majorities of interview texts also talked a lot about how the interviewees tried to work out structural difficulties, both in the design and the construction. The architects and other technicians needed to consider the methods to achieve the large span more carefully when designing the stadium projects.

Since Chinese architects formally participated in the design process of China-aided stadiums in the middle and late 1960s, its development fell a bit behind the development of Chinese domestic stadiums. Most China's foreign-aided stadiums (mainly outdoor stadiums), basically followed paradigms of Chinese domestic stadiums, but in lower standard and scales. Although Chinese architects felt a sense of freedom and possessed full discourse in the designing of China's foreign-aided stadiums, even facing the complex political environment in domestic China, it was not until the late 1970s that these overseas stadiums started to be dissimilar with the domestic ones in some aspects. Compared with other categories[①] of China's foreign-aided buildings, which tasted the fruits much earlier from Chinese architects' relatively unlimited creations, China's foreign-aided stadiums took more time from transferring domestic copies to unique designs. Therefore, generally in this period, the domestic architecture still had impacts on the designs of China's foreign-aided stadiums.

Climate aspects were considered in the designs of some China's foreign-aided stadiums of the 1970s. This can be regarded as the embryonic stage of regionalism in China's foreign-aided stadiums. The attempts were conducive mainly to natural ventilation and sunshine prevention, the development of which was prior to that in Chinese domestic stadiums. However, though the locations of these stadiums had a large cultural span from Asia to Africa, few cultural approaches were utilized in the architectural designs. Chinese architects tended to ignore the cultural elements in their free pursuing of modernity in these overseas projects.

In this period, as all China's foreign-aided construction projects were completely donor-oriented and referred to as "turn-key" project, the designs and constructions of which were charged by Chinese governmental departments and companies, so the recipient countries did not participate, nor did they express any opinions on the stadium gifts. These features gradually changed in later periods, with more involvements and opinions to affect the designs.

Generally speaking, Chinese architects concerned more about the function requirement, the architectural techniques, as well as the former stadium instance projects from domestic China. And some later projects cared about climate. This situation was gradually changing in the latter two stages (see Chapter 3 and Chapter 4, respectively), and the comparative analysis and summary of these vicissitudes will be explicated in the conclusion chapter.

---

① Chinese architects' works of other China-aided buildings shared unique characteristics since the early time, in the aid projects for Mongolia, including Choibalsan International Hotel (1960), Choibalsan Residence (1960), Ulan Bator Department Store (1961), in structuralist architecture, all designed by Gong Deshun from BIADI. And the institute also designed China-aided Sri Lanka's parliament building.

CHAPTER 3

THE 1980s TO 1990s — DEVELOPMENT PERIOD OF INDEPENDENT DESIGN

3.1 Transitional Period of the Reforming

In 1978, China opened its doors to the world and pursued more pragmatic policies. The famous "Reform and Opening-up" policy[①] started introducing market principles and gradually opened the Chinese economy to foreign investment and international trade. China shifted its focus on economic development and concentrated all its available resources thereon.

These changes not only positively promoted the development of China's foreign relations, but also directly affected the operation of China's foreign aid. China changed its idea of foreign aid and geared it towards the central task of development of its domestic economy. Since previous foreign-aid activities had placed enormous pressure on the Chinese economy, economic considerations became more influential in China's aid allocation decisions, whereby all diplomatic activities were supposed to gear towards the central task of the reforming period. While the scales of individual projects were reduced, mutually advantageous programs were promoted, with the grant element of Chinese aid fluctuating between 60% and 75% over the 1980 — 1985 period (OECD, 1987). Although China increased its aid substantially to Africa and Latin America countries in the 1990s (Taylor, 2006; Brautigam, 2008), the proportion of foreign aid to China's GNP(Gross National Product) was much lower than that of the previous period.

China's national foreign trade conference in 1980 changed the major idea of China's foreign aid, which proposed the new concept of "export and income, equality and mutual benefits". In January 1983, China's Leader proposed the four new principles of China's foreign aid in his visit to Tanzania, namely "equality and mutual benefit, being practical, diverse in forms and common development", which were formally put forward in China's sixth national foreign-aid conference that same year. The national conference on foreign economic relations and trade held in 1991 affirmed the principles to be the main guidance for China's foreign aid of the new period. Compared with the previous "Eight Principles", the "Four Principles" attached more focuses toward the economic benefits from aid and emphasized the two-way and commonality attributes.

China continued to provide aid to friendly countries but consciously cut the share of

① The "Reform and Opening-up" policy was proposed by Deng Xiaoping in 1978 to open the gate of economic mark of China to the world and to reform the society with more freedom and justice.

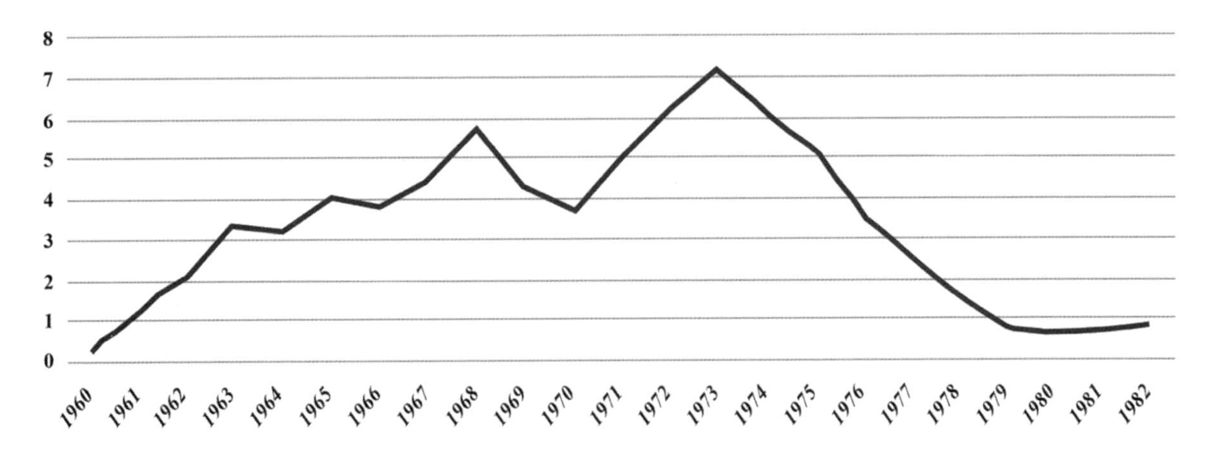

Fig. 3.1 Proportions of China's foreign aid of China's annual fiscal expenditure (1960-1982) (Source: drawn by the author based on the information from the National Economic Accounting Department of China's National Bureau of Statistics 1997)

aid to other countries. It also increased its foreign aid to Latin America, the Middle East and some African countries. In addition, China reoriented its assisting fields of foreign construction aid, reducing the number of industrial productive projects but increasing that of landmark buildings, such as cultural palaces, stadiums① and conference halls. Rather than undertaking large, costly projects, China preferred to fund small and medium-sized projects that were close to people's lives, such as agricultural demonstration bases, schools, hospitals, and necessary social infrastructures (Yu, 2016).

As China began to reform its economic system, the management of its foreign aid was simultaneously adjusted. The market mechanism was gradually introduced into China's foreign aid where the enterprise② became one of the principal parts. The ministry-responsible overall delivery mode of the planned economy period weighted the politics and diplomacy over economy, leading to a heavy burden on the country from foreign aid. In December 1980, China's MERFC (Ministry of Economic Relations with Foreign Countries) issued the "Interim Measures on the Trial of *Investment Lump Sum Contracting System*③ for Foreign Economic Aid Projects." Under the new system, Chinese government contracted the works

① The number of China's foreign-aided stadiums obviously increased, as explained in the following paragraphs.

② Many contracting units have been restructured from government departments to international economic and technological cooperation companies under governments at all levels, or other state-owned enterprises with legal status.

③ *Investment Lump Sum Contracting System* refers to the state subcontracting foreign-aid projects to a department that enjoyed a degree of flexibility and autonomy within the framework of legal policy (Liu, 2016).

of implementing foreign-aid projects to specific departments or the local governments, who took the overall economic and technical responsibilities and enjoyed autonomy in the project management (Zhang, 2006). This gradual elimination of the administrative orders authorized a certain degree of flexibility, turning the "governance" to the "management" for China's foreign aid projects, as seen especially in construction projects. The administrative departments in charge were able to seek enterprises that were willing to undertake the projects without interference from the Ministry of Construction.

With the deepening of China's economic reform, Chinese state organs began transforming their functions. In order to solve the problem of high cost and low efficiency in the *Investment Lump Sum Contracting System*, in December 1983, China's Ministry of Foreign Economic and Trade (MFET)[①] issued "Interim Measures on the *Contract Responsibility System*[②] for Foreign Economic Aid Projects", which allowed Chinese state-owned enterprises to bid for contractors positions, to share more decision-making powers, and to take on clearer responsibilities with regard to the construction of foreign aid projects (Shi, 1989). At the national foreign trade conference, China established the *Overall Contract Responsibility System*. China began to change its previous "turn-key" mode into foreign aid projects and participated more in the operation and management processes through market mechanisms to achieve greater efficiency and more benefits.

However, the system reform targeted construction enterprises over design institutes. In fact, since the late 1980s, tender and bid began to be experimented in the designs of China's domestic construction projects to meet the requirements of market system (Xue & Shi, 2003). Shanghai rang the bell first throughout the country to implement the tender and bid system in the designs of constructions, when, in 28 October 1988, its government released the "Shanghai Construction Engineering Bidding and Tendering Management Interim Measures. In 1992, Deng Xiaoping encouraged the formation of a market economy under socialist ideologies in his speech of Shenzhen Visit[③] (Xue & Shi, 2003). In the 1990s, larger construction projects in other Chinese cities also gradually entered the bidding process.

① In March 1982, the Ministry of Foreign Trade, MERFC, the State Administration of Import and Export, and the State Administration of Foreign Investment were merged to form the Ministry of Foreign Economic Relations and Trade (MFET). On 8 March 1982, MERFC was officially abolished.

② Under the *Contract Responsibility System*, administration and management were separated, introducing of the market system to improve the efficiency of implementation (Liu, 2016).

③ In September 1992, at the 14th national congress of the communist party, China formally announced that the goal of its economic reform is to establish a socialist market economy system.

However, this change did not spread to the designs of China-aided construction projects in the 1980s and 1990s. All the design work of these China-aided construction projects was directly assigned to the design institutes without a bid by MFET through COMPLANT, the company that was originally established under MFET. Except for the BIADI, more design institutes started being involved with the mission, including the core state-owned design institutes[①] located in major Chinese cities, such as in Beijing, Shanghai, Chongqing, Nanjing, Wuhan, and Shenyang. This new measurement consequently led to more diversified designs.

In 1993, the COMPLANT—under whose charge most China-aided stadiums were completed—was restructured as a comprehensive foreign trade enterprise group with independent accounting, self-management, and self-financing. The management of the aid projects was moved from the responsibility of "national ministries and commissions" to that of "enterprises' general contractors". Under these transitions, the stadium aid projects became more market-oriented and benefit-led. It required the architects to contribute more efforts to satisfy the recipient countries. The sense of competition also contributed to more consideration of the local contexts and requirements in the design of aid stadium projects.

Before the 1990s, China-aided projects were donated or built with donors or interest-free/low-interest loans. Even when loans became more common, interest-free grants were used to fund a considerable portion of early sports building aid projects. With this type of funding, the projects were limited to a certain scale— especially given that the Chinese government had been overwhelmed by increasing demands, following its transition from class struggle to economic development in particular(Table 3.1).

Table 3.1 Policy and mechanism of China's foreign-aided construction projects from the 1980s to 1990s

Period	1980s–1990s
Aid mode	Ministry-responsible mode; local-responsible mode
Principles	"[E]xport and income, equality and mutual benefits"; The new "four principles": "equality and mutual benefit, being practical, diverse in forms and common development"
Management department	China's Ministry of Foreign Economic and Trade (MFET); The local authorities of China; Other ministries of China; COMPLAN

① These design institutes were basically the provincial, municipal, and university design institutes.

continued

Period	1980s–1990s
Design institute	State-owned design institutes
Construction aid mode	Investment Lump Sum Contracting System; Contract Responsibility System; Overall Contract Responsibility System

3.2 Chinese Stadiums before and after the Beijing Asian Games

The economic reform in 1978 brought profound changes to China's society and economic development, as well opening the second developing-phase of Chinese stadiums. From the 1980s to 1990s, the relatively stable environment contributed to the significant progress of China's economic and cultural development. With the open door of China into the global world, Chinese stadiums became more diversified and modernized, having imported and applied new forms, functions, structures and techniques.

3.2.1 Promotion by National Sports Events and Asian Games

Since the 1980s, the sports events gradually became regular national activities in China, and the preparation of biding for the Asian Games was soon on the formal agenda[①], which stimulated the construction wave of Chinese stadiums. In 1984, for the first time, the team of China participated in the Olympic Games in Los Angeles. Chinese people excitedly witnessed the event through the newly popular black-and-white TV set. With the increasing scale of sports events and the improvement of the specifications, the construction scale and complexity of stadiums was greatly improved. This changed China's shortage of stadium construction in the early period, and numerous "sports complexes" emerged gradually in the form of a combination of high-standard stadiums and other sports venues.

Major Chinese cities started to construct their sports complex, such as Shenzhen,

① Early in 1984, Beijing formally submitted its Asian Games application to IOC and was granted the right to host the games, which was the largest sports event since the establishment of PRC.

Shanghai[①], and Guangzhou[②] (Fig. 3.2). China's national Olympic Sports Complex[③] was planned and constructed in Beijing for the Beijing Asian Games. In these sports complexes, the natatorium was normally designed and constructed along with the main stadium and indoor stadium. "One outdoor stadium, two indoor stadiums"[④] gradually became the standard configuration for sports complexes in China. Sports centers began to expand from first-tier cities to economically developed areas of China in the 1990s. And new types of stadiums were firstly introduced and constructed in China, such as the tennis stadium (Wang, 1990), skating stadium (Zhang, 1990) and other special venues for international sports events. The construction of these stadiums accumulated in major Chinese cities in the 1980s and 1990s, which established a solid foundation for further international competitions and urban development in China.

Fig. 3.2 Guangzhou Tianhe Sports Complex (Source: Ma, 2019)

3.2.2 Towards Modernist and Regionalist Designs

The opening-up policy permitted more communications with the global world. Academic materials[⑤] about sports architecture were introduced and compiled, and Chinese

① China's 1983 national games, held in Shanghai, was the first time that China hosted its national games in a city outside Beijing. The Shanghai Jiangwan Stadium became the main sports venue, where the newly built Shanghai natatorium improved the sports facilities of Shanghai.

② The 1987 Chinese National Games in Guangzhou was in fact a rehearsal for the holding of Beijing Asian Games. IOC President Juan Antonio Samaranch (1920-2010) was invited to attend the opening ceremony.

③ It includs a 20,000-seat stadium, a 6,000-seat natatorium, a 6,000-seat indoor stadium and other sports facilities.

④ "Two indoor stadiums" refers to one natatorium and one gymnasium.

⑤ The book *Foreign Architecture Examples – Sports Architecture* was edited and published in 1979 and the book *Art and Technique in Architecture* and the *New World Architecture (Sports Architecture* were translated and published in China.

architects had opportunities for architectural visit to other countries① before submitting their designs of Chinese stadiums for the Asian Games venues. For instance, the influence of the Yoyogi National Stadium can be seen in Ma Guoxin's② design of the roof with a hanging-cable structure in China's National Olympic Indoor Stadium. The unique structure of Beijing's Chaoyang Stadium, designed by Mei Jikui, was the result of optimization based on the roof structure of the Yoyogi National Natatorium in Japan, using the rigid cable net for the deficiency caused by using heavy steels, leading to the unique structure shape. Meanwhile, the "tender and bid" in the construction projects③ emerged in line with the transformation of China's economic mode. With the development of the "open competition" bidding mode (mainly for urban public buildings), foreign architects had the opportunities to step into China, consequently importing their architectural works. And the China–Japan Youth Exchange Swimming Center (1990) designed by Kisho Kurokawa became the first stadium designed by a foreign architect since the establishment of PRC. By the late 1990s, numbers of foreign architectural firms such as Canada's PDG, BDCI, America's JWDA and Japan's RIA, also designed stadiums in China(Fig. 3.3).

Fig. 3.3 Yoyogi National Stadium (left); China's National Olympic Indoor Stadium (middle); China-Japan Youth Exchange Swimming Center (right) (Source: http://m.sohu.com/a/211488443_711777; http://hansame.com/case_list/824.html; http://www.archcollege.com/archcollege/2018/6/40525.html?from=timeline&isappinstalled=0)

New attitudes and diverse architectural ideologies such as postmodernism, deconstruction, high-tech architecture, and critical regionalism were additionally pouring into China, which shook Chinese architects insofar as they were still attempting to fully digesting and

① For instance, architect Zhou Zhiliang, Mei Jikui, Wei Dunshan, and Ma Guoxin went to the United States, Canada and Japan in the 1984 Olympic delegation for stadiums to prepare the design of China's Asian games venues after they returned home.

② From 1981 to 1983, Ma Guoxin went to Japan for research and study in Kenzo Tange Architects, during which he had an intuitive and in-depth understanding of the Olympic venues in Tokyo.

③ In 1981, Jilin City of Jilin Province and Shenzhen City of Guangdong Province took on a new experiment in the project bidding, opening a new chapter of China's construction bid mode (Rao, 2008).

practicing modernism, and this undoubtedly caused the ideological turbulence in Chinese architecture. Chinese architects were no longer constrained by political ercironnent or the limitations created by nationalism, and thus began exploring new forms of modernism. For stadium architecture, volumes of voids and solids for pure images became one of the mainstreams. For example, different from the commonly used vertical segmentation style of facades in the 1970s' stadiums (e.g., Beijing Workers' Indoor Stadium), Sichuan North City Stadium (Fig. 3.4) chose to use a horizontal segmentation system, featuring a window belt in the middle to form the "solid" in the upper and lower parts and a "void" in the middle for contrast. The void from the long horizontal windows generates more sense of permeability compared with China's foreign-aided indoor stadiums in the previous period[①]. Another example, the Shenzhen Indoor Stadium (Fig. 3.5) adopted a simple quadrilateral plan, with the upper expanding stands intersecting with the roof with glass transitions and connections. Large areas of light-yellow walls and dark-brown glass deepened the contrast between the two dimensions, generating concise, clear and strong impressions (Shi, 2010). Chinese architects chose sculptural and volume shapes as the forms for stadiums, using large scale of beams, columns and surfaces to express the simplicity, innovation and power through the designs. Similar design approaches were utilized in other Chinese stadiums of the period.

More regional considerations were given in the designs of Chinese stadiums of this period. Some stadiums were designed to be adaptive to the local climate. For instance, in the main outdoor stadium of Guangzhou's Tianhe Sports Center, the edges of the stands overhung by 7 m, forming an open space as a rest platform from sunlight and rain, which were similar to the "qilou" (arcade) space commonly seen in Guangzhou area (Fig. 3.6). Some other stadiums expressed their regionalism in the shaping of culture and national image. This can be obviously found in the National Olympic Indoor Stadium, the design of which combined the slope roof image of Chinese traditional architecture with the modern sports building space and modern structures and techniques. Deformation of the traditional component "Dougong" were added into the architectural details of the cornices, and the multilayer stylobate also coincided with the large steps in the bottom parts like Chinese traditional palaces. These metaphors helped with the creation of modern sports buildings with Chinese characteristics. The Xishuangbanna[②] Indoor Stadium in Yunnan Province intro-

① Unfortunately, this stadium is under upgrading at present and will be changed completely.

② Xishuangbanna is an ethnic minority prefecture in China.

North City Indoor Stadium, 1980: (a) Main Facade of the Staduim; (b) Interior View of the Main Hall of North City Indoor Stadium, 2020:

(c) The Outside Facade under Upgrading

(d) The Rendering of the Upgrading

Fig. 3.4 Sichuan North City Stadium (Source: (a)(b) Ma, 2019; (c) https://m.sohu.com/a/193997440_713103; (d) shoot by Chen Yingting in 2020)

Fig. 3.5 Shenzhen Indoor Stadium (Source: the photo was taken by Sun Cong in 2019)

duced the sloping roof form and outer corridor of the "Zhulou" (bamboo house) to generate an iconic architectural image with local minority characteristics (Ding, 1986) (Fig. 3.6).

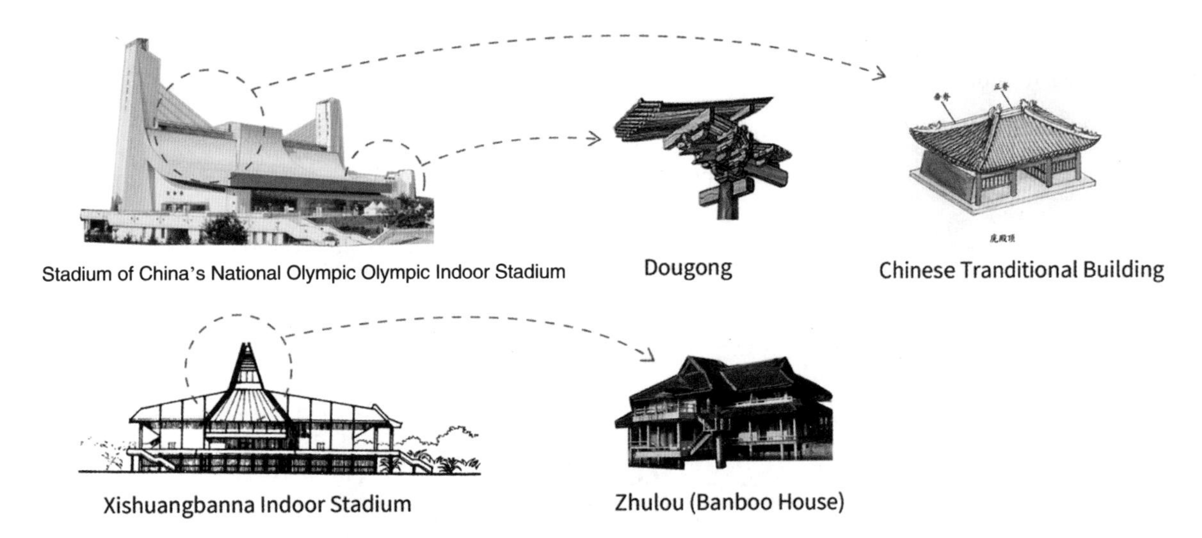

Fig. 3.6 China's National Olympic Indoor Stadium; and and Xishuangbanna Indoor Stadium (Source: drawn by the author)

3.2.3 Development of Architectural Techniques

Steel structures developed significantly and were widely used in Chinese stadiums during this period. Over 95% of newly built stadiums adopted a spatial grid or spatial truss structures (Sun, 1999). Based on the development of spatial structure, the layout and spatial language of stadiums had a wider range of freedom, forming a new sense of architectural features. Diversified forms of the reticulated shell structure were put into practice, and the twisted, extended combination or other deformation handling of the "arch shell" prototype enriched the architectural forms of stadiums, such as the Shanghai International Gymnastics Center and New Tianjin Indoor Stadium. New types of shell structures appeared such as the surface mesh shell, cylindrical mesh shell, twist mesh shell, and hyperbolic paraboloid mesh shell, which further enlarged the structure span from the approximately 100 m to nearly 200 m. Various types of cable structures[①] were experimented on in many medium or

① For example, the hub and spokes double-layer cable, hyperbolic paraboloid cable, cable truss plane cable system, spatial cable truss cable system, single-layer plane cable system, umbrella single-layer cable system, suspension cable network, suspension roof, and modular cable network roof.

large venues, leading to the development of cable membrane structure (Mei et al., 2002). In some stadiums, more than one type of structures had been combined to achieve new forms, such as National Olympic Indoor Stadium (1990) and Zhuhai Sports Center Natatorium (1999) (Zhou, 1996). In the mid- and late-1990s, the exploration of membrane structures contributed to more stadium forms. For example, the Shanghai Stadium adopted a large, cantilevered steel tube truss tension membrane, and Shanghai Hongkou Football Stadium (1999) innovatively utilized the cable membrane structure in its roof(Fig. 3.7).

Fig. 3.7 Shanghai Stadium (top); Shanghai Hongkou Football Stadium (bottom) (Source: top, Ma, 2019; bottom, photos were taken by Charlie Xue in 2019)

The overhang size and roof coverage area were also improved. Some large stadiums began to own a complete circle of the roof (e.g., Shenzhen Stadium [1993]① and Zhejiang Huanglong Stadium [1994]). At the same time, there were more options for the facades of stadiums in addition to vertical walls with matured curtain-wall technologies. The integration of the roof and wall become a new trend of coverage facades (e.g., the Heilongjiang Speed Skating Indoor Stadium [1995] and Shanghai's Pudong Natatorium [1999])(Fig. 3.8). Steel materials gradually replaced the concrete as the wall materials. The composite roof

① Shenzhen Stadium adopts the curved net frame (net shell) structure, which made the roof coverage area reach 20,000 m^2.

board made of high-performance alloy material can better meet the requirements of being waterproof, heat preservation and sound absorption for stadiums. The aluminum plate and other curtain-wall materials also became the important maintenance material of stadiums. Modern materials provided new opportunities for the architectural development of Chinese stadiums.

Fig. 3.8 Shenzhen Stadium (left); Zhejiang Huanglong Stadium (right) (Source: Ma, 2019)

3.2.4 Summary

Chinese stadiums of this period witnessed profound development. Under the influence of foreign architects' works, the significant development of structure and material, and the promotion from sports events, Chinese stadiums became more modernized and diversified. Chinese architects began to introspect more about the region, urban and internationalization. Chinese architects explored appropriate methods to combine tradition and modernity in stadiums. The "one main stadium, two indoor stadiums" layout and the core triangular homogeneous plan formed a typical Chinese sports complex, which had a far-reaching impact on the design of later sports centers. The following table 3.2 summarizes this development. These changes also affected the exportation output of China's foreign-aided stadiums to some extent, as will later be discussed in detail.

Table 3.2 Development of China's domestic stadiums from the 1980s to 1990s

Time	Scale (seats)	Architectural Features	Structural Development	Significant Event
1980s-1990s	More 60,000+ seat stadiums; More sports complex	Modernist and regionalist Designs; More diversified styles; Sports complexes with "one main stadium, two indoor stadiums" layout	Widely utilization of spatial grid or spatial truss structures; New forms of shell structures	China's National Games; Beijing Asian Games

3.3 China's Foreign-aided Stadiums: Conventionalized Designs with Basic Regional Concerns

In this period, China spread the stadium aid to a greater number of less-developed countries (LDCs) in Asia and Africa, and even extended its aid to Oceania and Latin America. In the 1980s, China supported the concept of South–South cooperation by boosting its aid activities in cooperation with developing countries. During this decade, over 20 stadiums were constructed, significantly surpassing the number achieved in the previous decade. Most of the outdoor stadiums were located in Africa, while indoor stadiums were more dispersed, including constructions in Africa, Asia, Oceania and Latin America. Although the average scales were around 30,000 to 50,000 seats for outdoor stadium and around 5,000 seats for indoor stadiums, similarly to the scales of the previous stadiums, more large-scale stadiums were exported, such as China-aided stadiums in Morocco, Zimbabwe, Kenya, and the DR Congo(Table 4.3). Compared with Chinese domestic stadiums, most of these oversea gifts were roughly equivalent to the sports facilities in Chinese medium-size cities. More stadium projects were designed in the 1980s than in the 1990s, especially large-scale ones, probably due to the constructions of stadiums in domestic China for the Beijing Asian Games.

3.3.1 The Gradual Forming of Conventional Designs

Similarly, the stadium aid during this period was usually proposed by the recipient countries, though not all were intended for holding sports events. Rather, some were for improving the local infrastructures of sports facilities generally. These China-aided stadiums became the first (or occasionally, the only) modern sports venues for the recipient regions.

During this period, Chinese local design institutes took the responsibilities for designing, and Chinese architects gradually formed the routines of economic and simple designs for these overseas projects. Compared with the previous stage, the designs of this period tended to be more cost-focused and aimed to satisfy basic functions.

Different from the domestic situation, in which numerous sports complexes were designed and constructed after the reforming and opening up, most of these China-aided projects were independent stadiums[①] rather than international sports complexes. Some stadiums were parts of the sports complexes of the recipient countries, which were sponsored and designed with the help of various donor countries or by the host countries themselves. Normally, China took the most difficult or costly mission, such as the main stadium, in cooperation with the local country or other countries. Chinese architects basically did not involve much in the general planning of the sports complex, and their designs of stadiums need to adapt to the exist site circumstances or designs. This had been a feature of China-aided stadium projects that persisted. For instance, the China-aided indoor stadium in Samoa in 1983 was part of the sports facilities of the Apia Park, which had long been in the process of design and construction. To help Samoa meet the requirements of the 7th South Pacific Games, to be held in September 1983, China designed and constructed a series of sports facilities in this country, including a 1,000-people gymnasium, a track field, a 2,000-seat stand, a bowling alley, and other facilities. These sports facilities, designed in 1980 and completed in August of 1983, gained much praise from athletes and referees from 19 countries during the Games (Ai, 1987). The designer was Ai Binggen from the Jiangsu Provincial Architectural Design Institute, one of China's state-owned design institutions(Fig. 3.9) And the Moi Sports Complex in Kenya was the exception (not only for this period, but also for the whole developing stages), and as one of the most significant stadium projects, it represents the relatively high design level of this period, which will be specifically analyzed in the following paragraphs.

China's Foreign-aided stadiums in the period tended to be the continuation of the previous stage, with a low development speed, a modest standard, a simple structure and an economic design. Most of the outdoor stadiums had single-layer stands, even with no independent seats, and the under-stand space was not utilized for additional rooms. Most of the stadiums did not have large-span full roof coverage, while a few of them had partial coverage of the rostrum areas, and some even had no coverage. However, it cannot be denied that

① The Moi International Sports Complex in Kenya was the only exception.

Table 3.3 Information of China's foreign-aided stadiums from the 1980s to 1990s

No.	Year	Continent	Country	Project Name	Design Institute	Capacity (seats)	Photo
Outdoor Stadium							
1	1980	Africa	Mauritania	Stade Olympique	East China Industrial Architectural Design Institute (Renamed as ARCPLUS)	10,000	
2	1980	Africa	Morocco	Stade Moulay Abdallah (in Labatt Sports Complex)	—	65,000	
3	1980	Africa	Gambia	Independence Stadium	—	17,000	
4	1981	Africa	Burkina Faso	August 4 Stadium	—	35,000	
5	1983	Africa	Niger	Seyni Kountche Stadium	—	30,000	
6	1984	Africa	Zimbabwe	National Sports Stadium	Gansu Architecture & Engineering Co., Ltd.	60,000	
7	1984	Africa	Senegal	One Stadium	—	50,000	
8	1985	Africa	Liberia	Samuel Kanyon Doe Sports Stadium (SKD Stadium)	—	35,000	
9	1986	Africa	Guinea-Bissau	One Stadium	—	15,000	
10	1986	Africa	Djibouti	El Hadj Hassan Gouled Aptidon Stadium	—	10,000	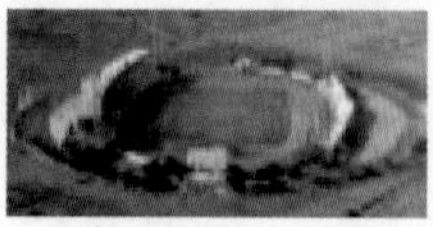

continued

No.	Year	Continent	Country	Project Name	Design Institute	Capacity (seats)	Photo
11	1987	Africa	Kenya	Moi International Sports Complex Stadium	China Southeast Architectural Design and Research Institute.	60,000	
12	1988	Africa	Rwanda	Stade Amahoro (Peace Stadium)	Design Institute of the Ministry of Railway (Renamed as Railway Engineering Consulting Group Co., Ltd.)	20,000	
13	1988	Africa	DR Congo	Stade des Martyrs (National Stadium)	China Southwest Architectural Design and Research Institute.	80,000	
14	1989	Africa	Senegal	Stade Mawade Wade	—	8,000	
15	1989	Africa	Mauritius	Stade Anjalay	China Sports International Co., Ltd.	18,000	
16	1989	Oceania	Papua New Guinea	Sir John Guise Stadium (National Sports Center)	—	30,000	
17	1990	Africa	Central African Republic	One stadium	—	30,000	—
18	1991	Africa	The Republic of Burundi	One stadium	—	20,000	—
19	1996	Africa	Congo	Massemba-Débat Stadium	—	17,500	
20	1996	Africa	Uganda	One stadium	—	40,000	—

continued

No.	Year	Continent	Country	Project Name	Design Institute	Capacity (seats)	Photo
21	1997	Africa	Mali	Stade du 26 Mars (National Stadium)	China Overseas Engineering Group Co., Ltd.	55,000	
22	1999	Africa	Togo	Stade de Kegue (National Stadium)	Beijing Institute of Architectural Design	30,000	
23	1999	Latin America	Grenada	National Cricket Stadium	China International Engineering Design & Consult Co., Ltd.	20,000	
Indoor Stadium							
1	1980	Oceania	Samoa	Indoor stadium in Apia Park	Jiangsu Provincial Architectural D&R Institute Ltd.	1,000	
2	1980	Africa	Morocco	Indoor stadium (in Labatt Sports Complex)	—	8,000	—
3	1982	Africa	Niger	Seyni Kountche Indoor Stadium	—	3000	
4	1985	Asia	Myanmar	Thuwunna Indoor Stadium	—	10,000	
5	1985	Africa	Tunisia	Natatorium (El Menzah Youth Sports Culture Center)	—	—	
6	1986	Asia	Yemen	National Indoor Stadium	Yunan Construction Engineering Co. Lt.d	—	

continued

No.	Year	Continent	Country	Project Name	Design Institute	Capacity (seats)	Photo
7	1986	Latin America	Surinam	Anthony Nesty Sporthal	Shanghai Civil Architectural Design Institute (Renamed as SIADR)	3,000	
8	1987	Africa	Kenya	Moi International Sports Center Gymnasium	China Southeast Architectural Design and Research Institute.	5,000	
9	1992	Latin America	Barbados	Garfield Sobers Indoor Stadium	Architectural Design and Research Institute of South East Univer-sity	5,000	
10	1994	Africa	The Arab Republic of Egypt	One Stadium	—	10,000	—
11	1994	Africa	Madagascar	Tananarive Indoor Stadium	Beijing Institute of Architectural Design	5,000	
12	1996	Africa	Sao Tome	One Stadium	—	10,000	

there are several large-scale stadiums (e.g., Moi International Sports Complex Stadium and the National Stadium of Congo) featuring two-layer or even three-layer stands with basically full coverage roof-coverage, which represented the improvement of Chinese stadiums exported overseas. These stadiums mainly intended for soccer games and track field competitions, with one special cricket stadium constructed in Grenada by the end of the 1990s. More categories of outdoor stadiums appeared in a later phase based on this beginning case, from which Chinese architects had an opportunity that had not been granted to them in the domestic Chinese context, namely the opportunity to design diversified functional stadiums.

Structures and construction techniques were no longer the serious problems that required by Chinese architects' solving, for there was no need of large-span roof structures for the stadiums, and the facades and forms were basically resembled the simple reinforced concrete frame constructions. "Nationalism" was totally replaced by modernism and functionalism, as illustrated by the similar appearance of China-aided stadiums in various re-

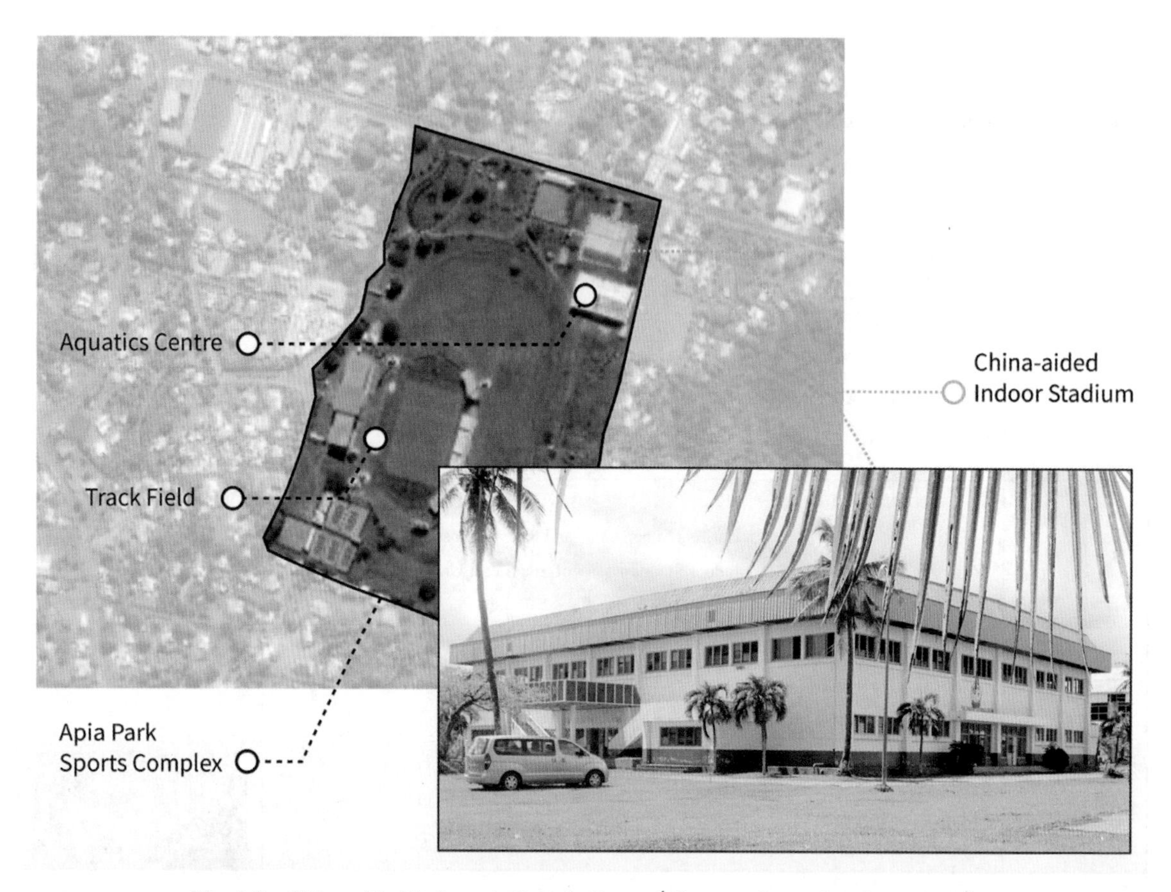

Fig. 3.9 China-aided indoor stadium in Samoa(Source: drawn by the author)

China's foreign-aided stadiums with similar features

August 4 Stadium in Burkina Faso　　Seyni Kountche Stadium in Niger　　National Stadium of DR Congo

China's foreign-aided stadiums with diversified features

Sir John Guise Stadium in Oceania　　El Hadj Hassan Gouled Aptidon Stadium in Djibouti　　Moi Stadium in Kenya

Fig. 3.10 China's foreign-aided stadiums with similar or diversified features (Source: drawn by the author).

gions in Africa or other areas (Fig. 3.11). The parallel forms, styles, materials, layouts, and structures were the results of conventional design exportation from Chinese architects, having spent decades working on these projects under economic limitations without complicated requirements. However, it should be noticed that there were still some stadium projects that had unique appearances and stood out from others, such as, among which regional expressions contributed to the differentiations. These dissimilarities were more obvious after the 2000s when diversified designs bloomed due to the changes in the policy, mechanism, architecture and time(see Chapter 4).

The spans of these oversea indoor stadiums were also controlled to be relatively small, amounting to no more than 80 m, whereas China's domestic indoor stadiums had surpassed the 100 m mark, and the roof structures were all basically the common plane grid steel structures. On occasion, the influence from the domestic projects could be found, such in

Fig. 3.11 Zhongshan Indoor Stadium in China (left); and Tananarive Indoor Stadium (right) (Source: http://archidatum.com/articles/in-africa-architecture-society-of-china_part-i/)

Fig. 3.12 China's foreign-aided indoor stadiums with regional features: (a) Thuwunna Indoor Stadium in Myanmar; (b) National Indoor Stadium in Yemen; (c) Seyni Kountche Indoor Stadium in Niger) (Source: http://cgcint.com/index.php/en/construction/others/139-zdgj/gongcheng/2014-06-24-13-58-24/329-niger5)

the China-aided stadium in Madagascar, which, coincidently, was highly similar to the Zhongshan Indoor Stadium in China (Fig. 3.12). However, some breakthroughs with special attentions in the designs could be seen in most of the China-aided indoor stadiums. Comparatively, the indoor stadiums located in various countries hold different architectural features, although their basic functions were alike(Fig. 3.13). Such differentiation was even more obvious than Chinese domestic stadiums of the same period, initiating the regional de-

Fig. 3.13 Sichuan Indoor Stadium: (a) View from the Main Entrance; (b) Design Model of the Stadium; (c) Façade of the Main Entrance (Source: (a)(b) were take by the author in 2023; (c) from Ma, 2019)

signs in China's foreign-aided stadiums beforehand.

Chinese architects explored modern design languages in most designs, with certain regional concerns in some ones. In most stadiums, low passive technologies were utilized in modern language to adapt to the local climate, including natural ventilation, natural light, or even the layout arrangement and hollow pattern on the facades for cost-efficiencies (Ai, 1987), which had also become parts of the design routines of Chinese architects, as will be explained in the following paragraphs about the interviews. Cultural aspects were tried to be expressed in some projects, such as the indoor stadiums of Myanmar, Yemen, and the famous Moi Sports Complex(see case study of this chapter) in simple methods with Chinese architects' own understandings. Some Chines architects wanted to experiment more and satisfied the local users better through the conventional working patterns and approaches.

3.3.2 Case Studies

For this period, two stadiums are chosen for the case studies, which are located in different continents: one in Africa and the other in Latin America. More importantly, the International Sports Complex in Kenya is one of the most well-known China-aided sports complexes overseas. It was designed by Chinese experienced architects, who had also been actively involved in various significant stadium projects in domestic China. For the China-aided indoor stadium in Barbados, it is one of the most modernized indoor stadiums that aided by China in the 1990s. The author believed that in-depth studies of these cases could well illustrate China's foreign-aided stadiums in the 1980s to 1990s.

3.3.2.1 Decorative Cultural Expression with Passive Low-technology: Moi International Sports Complex, Kenya

China and Kenya established diplomatic relations in 1963, experienced twists in the 1960s and 1970s, and returned to normal in the middle and late 1970s. In 1980, Kenya's president, Daniel Moi visited China for the first time and met with Deng Xiaoping. Moi requested China's aids in constructing international Olympic Game—standard stadiums in Nairobi for Kenya's application for the fourth All-Africa Games, which was actually promised by former Chinese Prime Minister Zhou Enlai. Deng agreed to donate the famous Moi International Sports Center and the journey of financial and technical assistance to Kenya thus began to recover.

Kenya was under British and French colonial rule for a long time, and Nairobi has attracted over 200 architectural design firms set up by the Westerners. Before Moi's visit to China, he sent people around the world to investigate about stadiums, and China's Sichuan Indoor Stadium (Fig. 3.13), designed by China Southwest Architectural Design Institute (CSWADI), satisfied the young leader. Therefore, he expressed his willingness to have CSWADI complete the design tasks of the Moi International Sports Complex. This situation was relatively unique throughout the 1950s to 1990s, since generally the recipient country did not have a decisive word about the aid stadium projects except for the scale and cost. This was basically the only case that the recipient country directly designated the design institute. Consequently, the sports center was designed by CSWADI with the chief designer of Sichuan Indoor Stadium Li Tuofen involved, and it was constructed by workers from Sichuan Province where CSWADI was located.

A design team led by Xu Shangzhi①, the chief architect of SCADI, together with Li Tuofen, Zhou Fangzhong, Sun Xianben, Zhang Jiarong, Shi Renyou and others, completed the planning and design works of the sports complex. Preliminary concept designs were first completed in domestic China before the field investigation, which had been a common routine for designing China's foreign-aided stadiums until the new century. In the winter of 1980, the design group went to Kenya to further improve the design and report to the local government. A team of nine people stayed in Nairobi for nine months. At the end of 1982, the design work was finally completed after 10 months' drawing back in China. And the blueprints for the buildings of the 80,000 m^2 gross floor area weighed thousands of kilograms.

In fact, at the beginning of the project, the design team had a debate about whether to transfer Chinese architectural features to the oversea aid projects or not. Architect Xu insisted on respecting the local culture and characteristics and enacted the final principles of avoiding the duplication of Chinese domestic stadiums abroad (Ping, 2000). However, when the team arrived in Nairobi, they found that after 500 years of colonial rule, the buildings in the city mostly mirrored the Western style. Chinese architects sensed the local identity from east African woodcarvings and the primitive huts② scattered across the countryside, which were believed to express the simplicity and roughness of America. How to integrate the traditional elements into the large-scale modern sports buildings became one of the main considerations for Chinese architects. They identified abstract architectural principles in Kenya's vernacular buildings and combined them with modern technology into the stadium (Xu, 2003).

The sports center is located in the Kasarari district of Nairobi, on the east side of Thika Road, the main road from Nairobi to the northern city of Kenya, and is about 13 kilometers from the CBD area. The site covered a large area of 100 hectares, and its planning was designed by Xu with garden-like layout, with a 60,000-seat stadium, a 5,000-seat indoor stadium, a swimming pool, and an embedded hostel for athletes. Rather than pursuing the symmetrical axis layout of the "one stadium, two pavilions" style commonly used in Chinese

① Xu Shangzhi was promoted to Chinese engineering design master in 1989 and he designed many significant constructions in China including the Chongqing Indoor Stadium.

② The primitive huts inspired Chinese architects in their designs of the athletes' hostel, which adopted the group courtyard layout and the hut style for the roof of the entrance.

domestic sports complexes, the planning of Moi International Sports Center basically followed the local topography and was in line with the traffic requirements. All architectures of the 80,000 m^2 gross floor areas were designed by Chinese architects. The outdoor stadium was mainly designed by architect Li Tuofen①, who designed most of the significant sports buildings in southwest China, including the Chengdu Chengbei Indoor Stadium, Sichuan Indoor Stadium, Chengdu Stadium, and later the China-aided Martyrs Stadium in the DR Congo. The indoor stadium was designed mainly by architect Zhou Fangzhong②. and Xu's colleague Sun Xianben designed the athletes' hostel(Fig. 3.14) after drawing inspiration from the huts(Fig. 3.15, Fig. 3.16).

Fig. 3.14 The athletes' hostel of the Moi International Sports Complex in Kenya (Source: *Collection of Chinese Excellent Architectural Designs* (5), 1999)

① Wu Defu, Zhang Jiarong, and Shi Renyou also participated in the design work of the outdoor stadium.

② Wu Defu, Wang Fuchun,and Li Ziyialso participated in the design work of the indoor stadium.

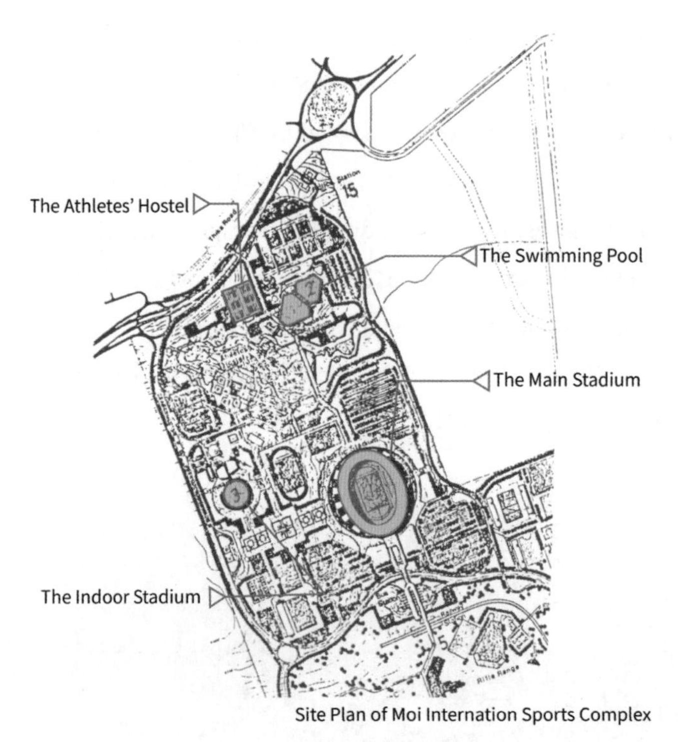

Fig. 3.15 Site plan of Moi International Sports Complex. (Source: *Collection of Chinese Excellent Architectural Designs* (5), 1999)

Fig. 3.16 Designs of the site layout of Moi International Sports Complex; and models of Moi International Sports Complex (Source: provided by CSWADI)

The main stadium features 54,000 m^2 of gross floor areas and 60,000 seats. Its layout was divided into 24 sections, each taking on a petalage outline with a tilted-column-supporting cantilevered stand. The blossom image was inspired by the local plants in the site photos provided to Chinese architects before their visits and later was described to symbolize the friendship between China and Kenya. Due to the economic limitation, its structure remained a reinforced concrete frame with double arrays of columns, as was commonly used in China's foreign-aided stadiums of the 1970s. However, the three-layer stands and the partial coverage of the full circle of the platform showed much improvement in comparison to most other aid stadiums that only hold a one-layer stand without coverage. In considerations of the local climate, low passive technologies were utilized in the design of the stadium to save costs. The spaces between the adjacent petal parts formed large holes between the stands and the roof to introduce more natural ventilation while spreading the sound of the lively competition to the outside. In addition, hollow space grid windows, each made of a concrete frame, were widely used as the facade of the main entrance hall, low-floor rooms and other interior transport space, which effectively balanced the sunshade and ventilation required by the local climate. The open lounge and internal courtyard were adapted to make the space treatment of the building transparent and vigorous. Such low passive technologies were commonly used in China's foreign-aided stadiums, especially in Africa or other tropical regions where natural ventilation outweighs mechanical ventilation (Fig. 3.17).

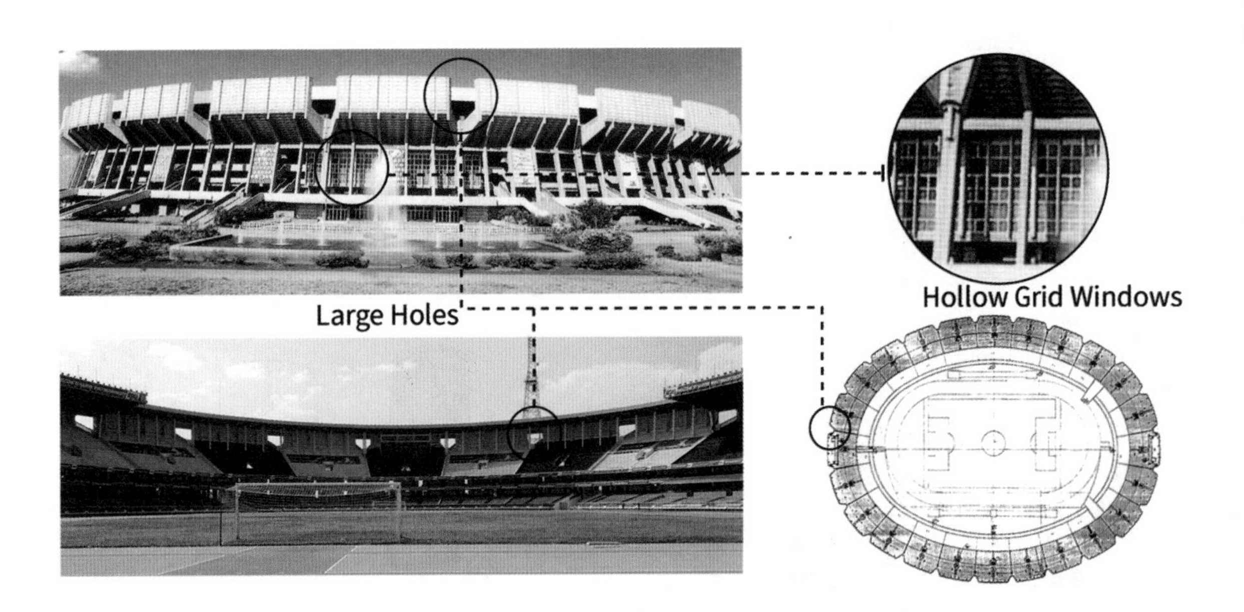

Fig. 3.17 Regional design approaches of the stadium of Moi International Sports Center, Kenya (Source: redrawn by the author based on pictures from *Collection of Chinese Excellent Architectural Designs* (5), 1999)

In the visits to Kenya, the architect discovered that the traditional African woodcarvings were clearly hung obviously in the living rooms of houses and embassies, seen as the most valuable gift to the most honorable guests (Fig. 4.18). Chinese architects regarded it as good expressions of the local culture. Concrete relief sculptures based on the woodcarvings were cladded above each entrance both as decorations and deformation joints between the double columns. Chinese architects could only illustrate the local identity culture in stadiums through such easily decorative components, probably due to the economic limitation, architectural, structural and material development. Exposed post and beam structures were believed to symbolize the Maasai spirit of East Africa. The special pattern formed by the concrete construction technique also sought to echo the local style. It was achieved by Chinese laborers with high crafting skills, which may fail to pass down to future generations. Such craftsmanship can be rarely seen in Chinese contemporary concrete buildings. Chinese technicians conveyed their understandings and expressions of the local culture in economical methods.

The stadium was designed to be a modern stadium capable of holding international soccer games and track competitions, featuring facilities such as playing fields and tracks, an electronic timing scoreboard, sound amplifiers, telephones, telecommunications and lighting, as well as other additional facilities such as the president room, VIP rooms, audience lounges, a bar, a square, and parking areas(Fig. 3.19).

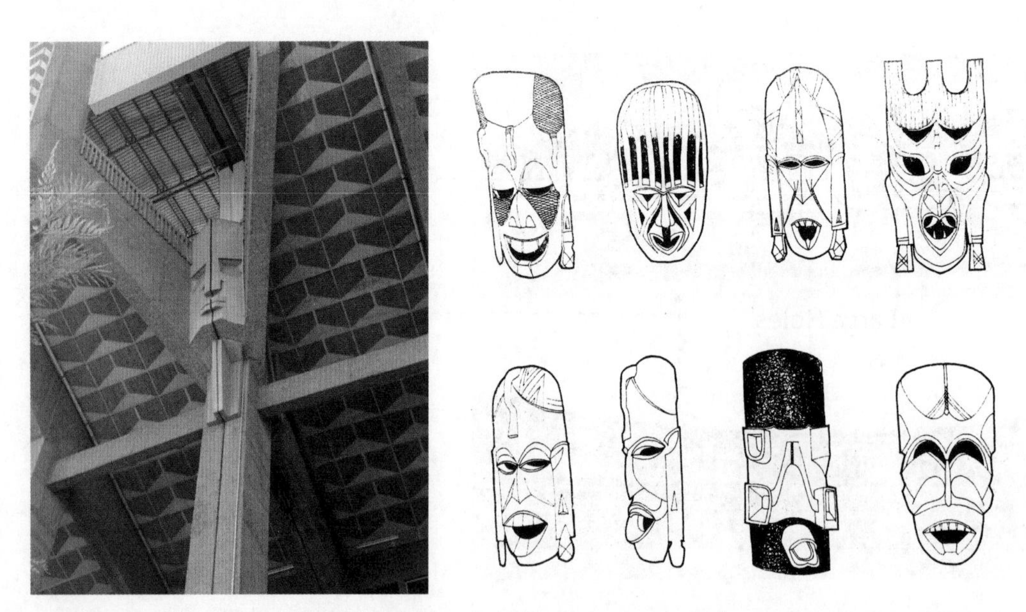

Fig. 3.18 The concrete sculptures (left) and the woodcarvings in Kenya (right) (Source: left, the photo was taken by Wang Daozheng; right, from Zhou, 1984)

Fig. 3.19 The inside views of the stadium of Moi International Sports Complex (Source: the photos were taken by Wang Daozheng in 2018)

The indoor stadium has a gross floor area of 12,250 m^2 which can hold 5,000 spectators. Its layout also adopted the petal-shaped plane (an octagon). On the premise of not obstructing the audiences' views, the eight large supporting columns were aligned inward to reduce the roof span to under 70 m (compared with the domestic span). The audience area was thereby divided into 8 sections, making it easier for the evacuation and reducing the interference between different audience areas. The petal-shaped plane also increased the areas of side windows of the competition hall to improve the interior lighting environment, which enabled general trainings and even some competition activities with natural lighting. The main entrance was on the second floor, equipped with an outdoor gallery, and led to the underground through four single-run steps and a slope. Four entrances to the auditorium were located around the inner corridor of the second floor, which were staggered from the outdoor entrances to form an orderly traffic space. The auditorium was color-coded, and the terraced auditorium gave a panoramic view of the game on the ground floor. The competition ground and the audience lounge were at the same height so that the under-space of the first-layer stand could be used for additional rooms. The audience lounge area also happened to make full use of the understand space behind the crosswise walkway, such that nearly all of the understand space could be used. The design tried to meet the basic function requirements with the most compact and economical layouts(Fig. 3.20).

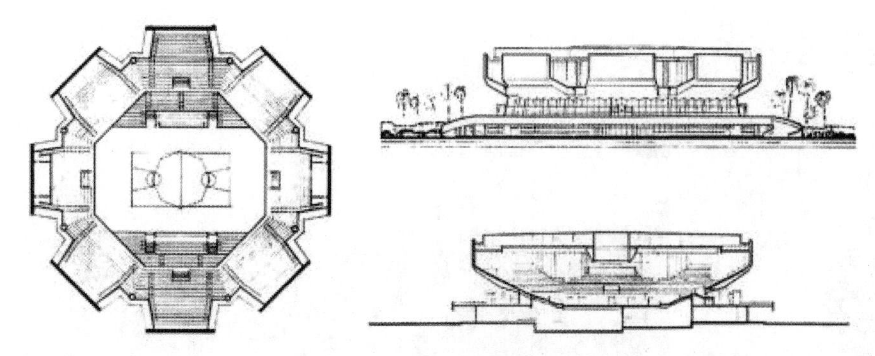

Fig. 3.20 Indoor stadium of Moi International Sports Complex (layout, facade, and section) (Source: CSWADI, 2015)

Interestingly, although President Moi favored the Sichuan Indoor Stadium, the roof of the indoor stadium roof did not use a complex curved-shaped structure such as that of the Sichuan Indoor Stadium did, but instead of using the simple plane grid system, which was more economical and practical. Perhaps the designs were made in consideration of economic limitations and the difficulties of overseas constructions. The construction was completed through the jacking technologies again, developed from the jacking of four columns to that of eight columns(Fig. 3.21).

Fig. 3.21 Indoor stadium of Moi International Sports Complex (outside and inside views) (Source: the photos were taken by Huang Zhengli)

In fact, the Sichuan Indoor Stadium that President Moi favored also had a relatively unique-shape layout rather than the normal square one. Zhou chose the petal shape for the indoor stadium as a continuity of the outdoor stadium, being mainly concerns about the basic functional, technical and economic aspects of the design. The petal-shape reflected the internal functions, but right also coincidentally represent the energetic character of the sports buildings and show a certain lively and romantic spirit of the African wilderness. There were no sculptures on the facade or entrance, nor was the facade concrete made with a pattern but directly exposed. A totem artwork was hung upon the stairs in the lobby facing the main entrance. It is worth mentioning that decorations of the lounge, VIP rooms, and other parts of the interior adopted the classic Chinese style, which was believed to be a compromise with the different opinions from the design group before the designing(Fig. 3.22).

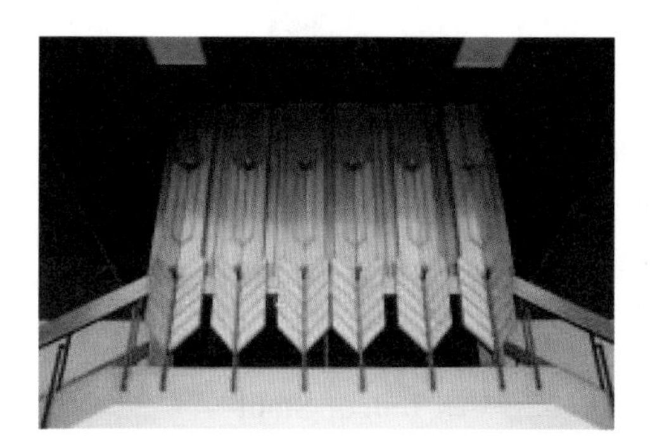

Fig. 3.22 Decorations of the indoor stadium of Moi International Sports Complex (Source: right, taken by Huang Zhengli; left, taken by Wang Daozheng in 2018)

The Moi International Sports Center finally melt the clients' doubt (Xu, 2003), and satisfied President Moi. The fourth All-African Games was held there in August 1987, soon after its completion. IOC President Antonio Samaranch presided over the opening ceremony with 110,000 people taking part in the events. The sports center functioned well and the IOC president was impressed with the Chinese design and construction. To commemorate the project, Kenya printed the main stadium of the Moi International Sports Centre in its 20-shilling note(Fig. 3.23). China continued to provide maintenance assistance and technical cooperation to the sports center, which helped with its good conditions and continuous use later on. Furthermore, CSWADI started to design additional China-aided stadium projects overseas such as stadiums in the DR Congo and Ethiopia.

Fig. 3.23 Kenya's 20-shilling note with Moi Stadium (left);and a report about the fourth All-African Games in newspaper (right). (Source: https://www.ghettoradio.co.ke/how-moi-plunged-stadium-into-darkness-to-hand-harambee-stars-a-win/;https://kenyapage.net/commentary/kenya-football-articles-and-profiles/kenya-at-the-4th-all-africa-games/)

3.3.2.2 Modernism with Economical Regional Concerns: China-aided Indoor Stadium in Barbados

China and Barbados officially established diplomatic relations in 1977, making Barbados one of the very first English-speaking Caribbean countries to forge diplomatic ties with China.[①] Barbados and China have been working together since then based on mutual trust, cooperation and shared values. It was after the 1980s that more regions besides Asia and Africa began to receive China's stadium aids, such as the Oceania and Latin America, among which the China-aided indoor stadium in Barbados is a representative case.

The stadium project was the largest economic aid project undertaken by China in the eastern Caribbean. Initially, according to the Barbados law, the government was not entitled to receive economic assistance through bilateral channels, which delayed the process of the project for some time. Then its parliament passed the "amendment to the law on special payments" to allow the acceptance of foreign aid through local channels. On 1 March 1988, the agreement between China and Barbados on the stadium project was approved by Barbados' parliament, and the design contract[②] was signed later. For a scenic island nation in the eastern Caribbean, Barbados expected the stadium to promote both its sports and tourism development.

The stadium was financed by an interest-free loan, with a total investment of RMB 33.33 million yuan, which was assigned to Jiangsu Province as one foreign aid mission by

① http://bb.china-embassy.org/chn/zbgx/t1265647.htm.

② The project contract was signed by the minister of tourism and sports of Barbados, Mr. Wyse hall, and the Chinese ambassador of Barbados, Mr. Lu Zongqing. (Fang, 1988).

MFET. Therefore, both the design and construction organizations were from Jiangsu. The design work began in 1989 and the construction was completed in June 1992. The secretary-general of China's state council, Luo Gan, handed over the stadium to the government of Barbados on behalf of the Chinese government and Cao Keming, the deputy party secretary of Jiangsu Province was invited to attend the opening ceremony (the General Office of Jiangsu Provincial People's Government, 1993). The stadium was praised as "one shining pearl of the Caribbean" by one of the popular newspapers of Barbados, the *Barbados Advocate* to express their satisfaction.

In fact, the indoor stadium was the main construction of the Sir Garfield Sobers Sports Complex situated in Wildey, St. Michael, Barbados. The complex was part of a strategic initiative by the government of Barbados to promote sports tourism, which houses the China-aided indoor stadium (Wildey Gymnasium), an aquatic centre, tennis courts, an exercise room, a medical bay, warm up rooms, and the Wildey Turf football stadium (Fig. 3.24). Surrounded by high grounds, the site of the sports complex lowered gradually from northeast to southwest, and the indoor stadium was located in the southwest where the terrain was relatively flat. To donate one or two sports constructions of the sports complex—normally the most pricy or complex ones—was one of the routines of China's foreign aid activities, like the Wildey Gymnasium being the focused attraction for the newly constructed sports complex. It was the first time that China's foreign-aided stadium projects were designed by a design institute of a Chinese university, since the Southeast University had one of the best architectural schools in China and its design institute stood for the high levels of Chinese domestic design institutes.

The university's architectural design institute seemed to be able to carry out modernism more simply, directly and academically. Chief architect Yang Weihua used a regular central axis layout in the 16,000 m^2 area of a site for which China was responsible for. The central axis square faced the main entrance to the indoor stadium directly without any obstacles, flanked by the tennis court and the swimming pool. The pedestrian part and the vehicular part were separated along the diagonal of the latter one, and the parking areas occupied half the area of the site to meet the needs of the stadium by providing a space for various activities. Following a rigorous geometry, the square site was centralized by the square indoor stadium (Fig. 3.26).

The 4,000-seat indoor stadium covered 9,941m^2 of gross floor area in a 66 m × 66 m square-shape plan (Yang, 1994). The main entrance for the audiences was located in the east with the exit in the west. Athletes, performers, VIPs, and staffs had their own entrances without interference. The standard basketball ground in the center was surrounded by four-side stands, of which the east and west were three-stories while the north and south were one-story. Most of the internally used additional rooms were on the ground floor, and the audience-used ones were arranged on the north and south sides of the first floor. The neat and compact layout emphasized on economy and basic functions, as did most other China-aided stadiums of the period (Fig. 3.26). The roof adopted a relatively simple structure, namely a plane grid system supported by eight columns with the span controlled within 70 meters, and a wall bearing structure systems filled in the other parts of the stadium.

The lifestyle in Barbados tended to be modernized and Western due to its being one of the popular resorts of the Caribbean. Chinese architect chose to use more purely modernist architectural language in the stadium of the island. The facade walls used numerous small white bricks as the background of the local context and were intended to reflect the modern features. Two staircases on both sides ascended towards the main entrance of the stadium simultaneously, and the resting space between used lattice bars as the shading components,

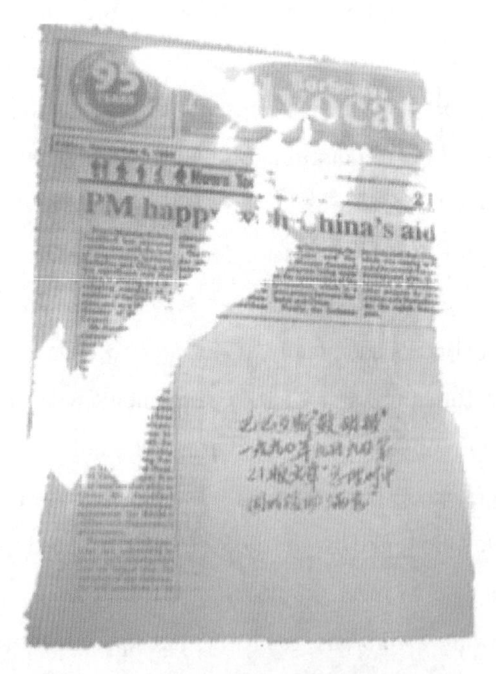
PM happy
China's aid

Fig. 3.24 The report about the China-aided indoor stadium in the Barbados *Advocate* journal, 1990. (Source: http://kaifangzhan.mofcom.gov.cn/article/g/i/200902/20090206026723.shtml, achieved in 2018)

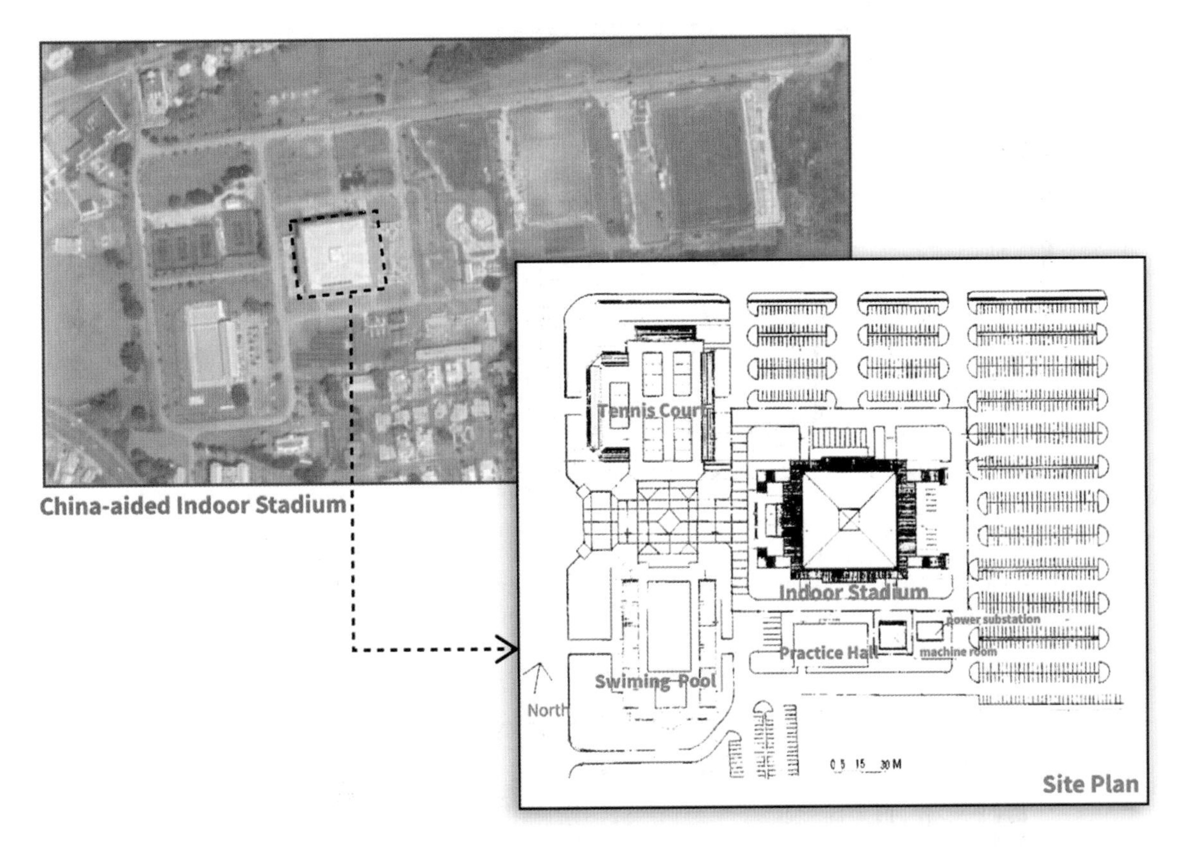

Fig. 3.25 Location of Sir Garfield Sobers Sports Complex; and the site plan of the China-aided indoor stadium in Barbados (Source: drawn by the author)

forming the horizontal lines in the central composition to contrast with the large solid areas of the facade. Powerful vertical structural components and a horizontal cantilevered auditorium were combined to form a modern-style indoor stadium (Fig. 3.27).

Yang also wanted to combine certain traditional elements into the modern block and noticed the roof characteristics of Barbados' traditional chattel house, the red sloping roof. During the time, Chinese architects seemed to conventionally consider roof elements as good expressions of local culture, as they did in many Chinese domestic buildings, especially the sports buildings. Yang chose to use corrugated red metal planes to cover the partial sloping areas of the roof of the stadium, reflecting the traditional and local characteristics. This sloping shape of the roof was formed by removing the outermost edge of the upper strings of the grid frame and keeping the surrounding diagonal strings. This approach did not actually change the roof's grid frame structure much, which is still regular, nor did it

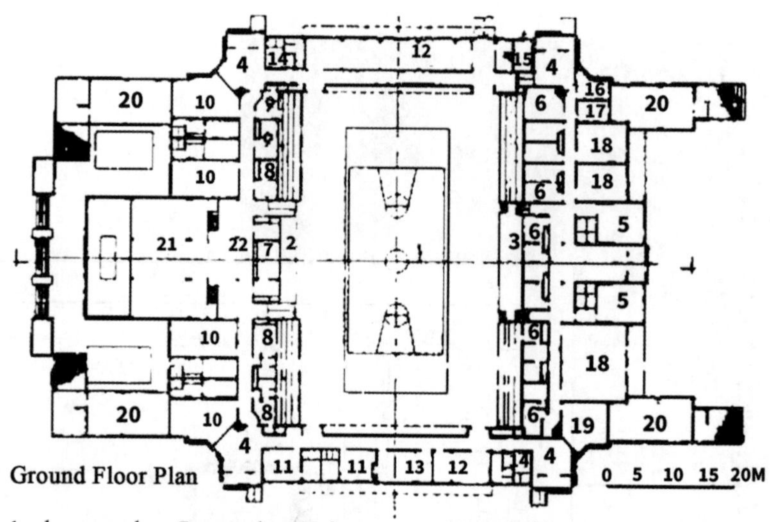

1. playground
2. judge seats
3. rostrum
4. entrance hall
5. VIP lobby
6. office
7. central control room
8. medical room
9. broadcasting room
10. lounge for etheltes
11. lounge for judges
12. sports equipment room
13. switching room
14. toliet
15. guard room
16. ticket room
17. control room
18. meeting room
19. reporter station
20. air-conditioning control room
21. fitness room
22. video hall

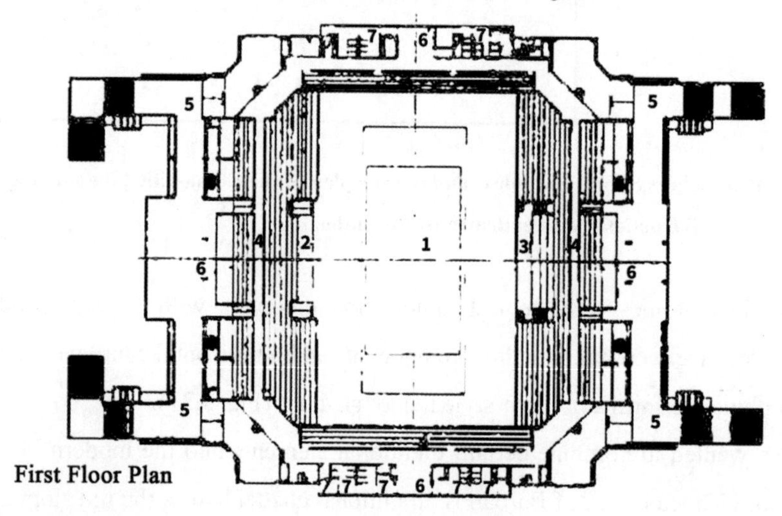

1. playground
2. judge seats
3. rostrum
4. audience area
5. entrance and exit for the audience
6. drinking area
7. toliet
8. shops

Fig. 3.26 Floor layouts of the China-aided indoor stadium in Barbados (Source: Yang, 1994)

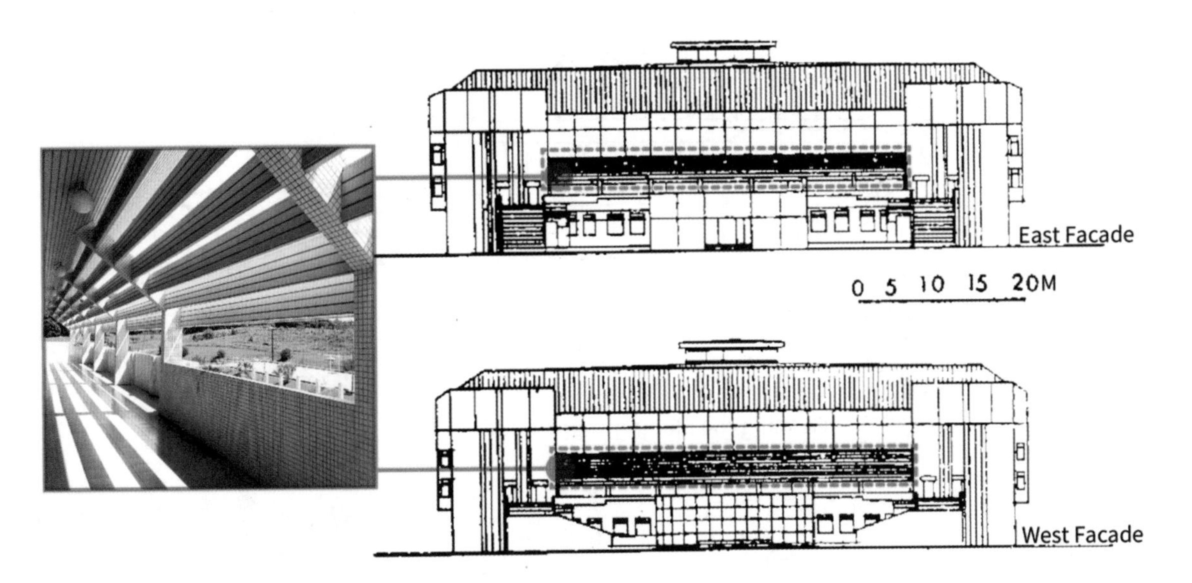

Fig. 3.27 The entrance design of the China-aided indoor stadium in Barbados (Source: Yang, 1994; https://www.picuki.com/media/2138155101884466504)

impose extra burdens on the structure or in terms of cost. The architect combined local features in his modernist work in a simple and economical way, through which visitors had views of a bright-color facade with red sloping roofs, clean walls and modernized forms (Fig. 3.28).

It can be easily observed that China-aided indoor stadium in Barbados interestingly shares similarities with the indoor stadium in Samoa (see above sections), probably due to the fact that both architects considered the local features and environments of these two coastal countries. This is in contrast to the Moi Indoor Stadium in Kenya. Compared to the stadium in Kenya, the space of the Barbados indoor stadium was more closed and its facades more solid. The interior relied more on artificial lighting and air conditioning systems, likely as a result of the more comfortable climate and better infrastructural development of Barbados. At the same time, the closed infield environment providesd better support for more categories of performances that the stadium could hold(Fig. 3.29).

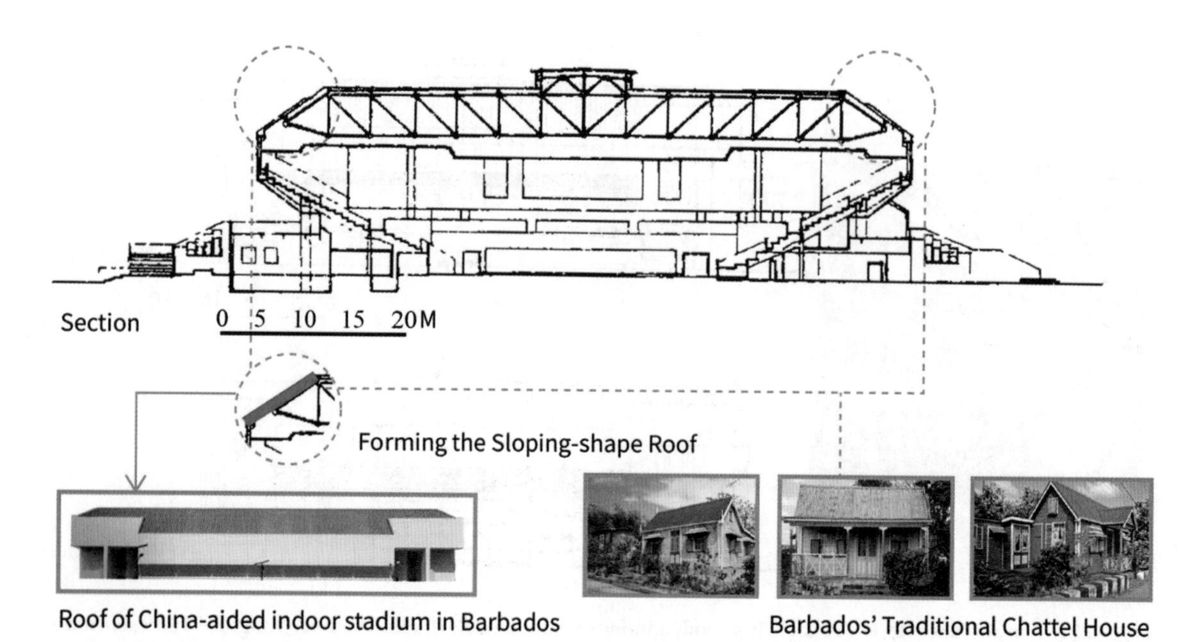

Fig. 3.28 Roof form designs in the China-aided indoor stadium in Barbados (Source: drawn by the author)

After the completion of its construction in 1992, this China-aided stadium was one of the best modern indoor stadiums in the Caribbean in the 1990s. The Wildey Gymnasium, as it is commonly known — had become Barbados' favored venue for hosting musical concerts, as well as religious, cultural, and of course, sporting events, among other events.

Fig. 3.29 The entrance facade (top); and the outside view (bottom) of the China-aided indoor stadium in Barbados (Source: from CSIG)

3.4 Summary: What Stayed and What Changed on Chinese Architects' Tables?

The 1980s—1990s period set the historic stage of China's economic reforming and also the second stage of the development of China's foreign-aided stadiums overseas. The number of the stadiums increased substantially and the geographic distribution of the stadiums spread from neighboring Asian countries and some African countries to most African countries (the greatest number there), as well as other continents such as the Oceania and Latin America.

Chinese architects used smaller scale, smaller span, and compact functional layouts in the designs of these stadiums. Additionally, low passive technologies, simple structures, and cheap materials were encouraged in the design of China's foreign-aided stadiums in accordance with the climatic aspects like natural lighting and ventilations, partly due to the low infrastructure development levels of the recipient countries, and their incapability of maintenance, and more importantly, to limit the overall cost of the projects.

From the 1980s to 1990s, the large-span structures of China's domestic stadiums developed greatly with the construction wave of high-standard stadiums of the period. Therefore, when designing oversea-aid stadiums, the structure was no longer one of the main difficulties and focuses. Moreover, under strict economic control, these stadiums basically used simple plane structures and basic reinforced concrete structures that could be easily implemented based on domestic experience, and to the contrary of domestic projects, new structural forms were not experimented with. Similarly, materials, construction and other aspects of architectural techniques tended to be excluded as the limitations—especially for foreign stadiums, which used economic materials and construction methods.

As required by the recipient countries, some of China's foreign-aided stadiums were still designed and constructed for hosting regional or international sporting events, while others were intended to promote urban development and improve sports facilities for residents. Some specialized categories of stadiums such as natatoriums and cricket stadiums started to be exported, which demanded the professional transitions of oversea-aid stadiums. Moreover, some venues even had to accommodate post-game operations, such as events and performance holding. Chinese designers needed to design according to these various requirements as well as the basic functions. During this period, the diversified functions improved the aid stadiums and were prepared for more specialized ones in the new era.

As China's domestic architecture walked towards freedom without restrictions, so did Chinese architects when they designed stadium projects, both at home and abroad. The trend of appealing for "architectural creation" encouraged architects to consider the design features and form conventional routines in designing of foreign-aided stadiums. Since most of the oversea stadiums featured smaller scales and lower standards compared with domestic ones, many of them looked similarly to China's conventional, domestic stadiums in the 1970s and 1980s. Of course, several significant large stadium projects (or sports complexes) spoke for the specificity and regionalism.

At this stage, climate fundamentals were considered in most projects, especially for stadiums in Africa, in which natural ventilation and natural lighting became the necessary factors to consider in the designs. Most low passive technologies adopted were also for climatic effects, and more attention was paid to this than in the previous stage. In terms of culture, some Chinese architects tried to identify approaches to combine regional culture in modern stadium designs. They preferred to use simple and economical methods, such as decorations and materials, by understanding and expressing the cultural characteristics.

Generally speaking, from the 1980s to 1990s, Chinese architects were more concerned about the economic and climate aspects in their designs of China's foreign-aided stadium projects. This can explain the obvious transitions of these stadium projects to become economical and climate-concerned buildings with low passive technologies, simple structures, and cheap materials, compared with China's domestic stadiums. Most of China's foreign-aided stadium projects did not surpass the domestic ones in scales, standard, or other architectural levels. Therefore, the development of Chinese domestic stadiums could easily support the oversea ones without much restriction. It should be noticed that with the increasing climate and culture considerations, the regional factors tended to exert more influence, which would be more obvious and become the mainstream in the later period (see Chapter 4).

The author regards the technologies and design approaches utilized by Chinese architects as adaptive technologies and approaches for the good cost performance and effectiveness of these oversea projects under special circumstances and limitations. At the same time, some of the significant projects (e.g., the case study) witnessed the growing up of Chinese architects in design, especially in their regional design attempt. However, there is no denying that, in general, there are fewer high-quality works in this stage than the previous stage, as confirmed by some of the interviewees' statements. The design level was unremarkable and the innovation was defective, which likely made these architectural exportations contribute less to the improvement of Chinese stadium architecture probably in this period. This situation was changed in the new millennium era, when the similarities and dissimilarities intersected with each other. (see Chapter 4).

CHAPTER 4

THE 2000s TO 2010s — CRITICAL REGIONALISM WITH SIGNIFICANT IMPROVEMENT

4.1 Towards Market Reform and Bilateral Mechanism

In the new century, as some Western governments scale back their development finance commitments, non-Western donors are rapidly expanding their overseas aid activities. China has rapidly expanded its overseas aid activities, becoming one of the main non-Western donors (Strange at al., 2017). In 2000, China launched a joint initiative with 50 African countries and promoted the establishment of a multilateral assistance mechanism known as the China Africa Cooperation Forum (CACF), to be held once every three years. For the first time, China proposed reducing and exempting the debts of the heavily indebted developing countries and the least developed countries and continuously expanded its assistance to these underdeveloped countries (Yu, 2016). In 2006, China declared a "new strategic partnership" at the Beijing Summit of the Forum on China—Africa Cooperation (FOCAC), announcing plans to double its 2006 aid effort to Africa by 2009 "to reach the target of mutual benefit and the win-win situation between China and African countries" (Ministry of Commerce, 2004-2009). Under "The Silk Road Economic Belt and the 21st-century Maritime Silk Road" Initiative (One Belt and One Road Initiative, BRI) proposed by China's leader Xi Jinping in 2013, China's aid policy entered a new era. China's aid began to favour the countries along the BRI, mainly including nodes countries in Asia, Africa and Europe[①]. China increased its assistance to the livelihood and public welfare undertakings in the third world countries such as medical, educational and sports (with stadiums included) aids, etc., to enhance the economic and social capacity of the recipient countries and improve the quality of their people's lives (The State Council of China, 2011; Liu, 2016). During the new period, China saw the achievement of mutual economic development as the priority and still regarded infrastructure constructions as the main assistance projects (Liu, 2016).

The tide of marketization promoted the reform of China's foreign aid policy to enter a completely different period (Kobayashi, 2008). Since the mid-late 1990s, the Chinese government planned to introduce the market to function more in line with its domestic economic reform and the "going out" strategy. Since 2000, more forms of economic aid were im-

① The nodes countries mainly include Mongolia, Russian, 11 countries in the Southeast Asian, 8 countries in the South Asian, 16 countries in the West Asian and North African, 16 countries in Middle and East Europe, and 5 countries in Middle Asian, (https://baike.baidu.com/item/%E4%B8%80%E5%B8%A6%E4%B8%80%E8%B7%AF/13132427)

plemented besides grants, such as the interest-free loans and concessional loans. Grants are generally used more in the LDCs, whereas loans are often a larger portion of the aid portfolio in more developed countries (Hubbard, 2017), which led to the diversification of the main bodies of China's foreign aid. Some Chinese state-owned financial institutions such as banks became an important source of aid funds and one of the main bodies, while some Chinese large state-owned enterprises and renowned NGOs gradually participated in China's foreign aid in some specific aid projects①, which changed the situation of the Chinese government being the only body in aid activities. The involvement of financial institutions introduced market funds into the field of foreign aid, helped to mobilize the funds from both the donor and the recipient country, promoted the development of the recipient country through cooperation in between enterprises and financial originations of the two sides, and inspired more cooperation in other aspects or fields of the two countries. All of these made China perfectly from a single donor to the mutual benefit of both parties. China's aid diplomacy transformed from generous grants to cooperative agreements.

Under such new financial supporting mode, the aid programs are supposed to be collaborated with various financial institutions with an adaptive aid package according to the development needs. For construction aid projects, the government-led free donor (completed aid projects) and financial institutions-led concessional loan cooperation become two major assisting methods, among which the latter providing a larger and more diversified financial support for construction projects. This also contributed to the profound improvement of China's foreign-aided stadiums of the new era, which right unlikely happen without better financial supports.

To adapt to the accession to the WTO②, China began to reform its administrative system. In March 2003, the Ministry of Commerce of the People's Republic of China was established by integrating the functions of the former MFET with other governmental departments③. The Department of Foreign Assistance of MOFCOM was responsible for formulating regulations related to foreign aid and the main management of foreign aid projects. At the same time, the Agency for International Economic Cooperation (AIECO)of MOFCOM was responsible for the assistant manager of foreign aid affairs. After decades'

① For example, after the Indian Ocean earthquake and tsunami disaster in 2004, the Red Cross Society of China donated RMB 260,000 yuan, and the China Charity Federation donated RMB 23,794,230 yuan. (Pang et al., 2005).

② China entranced the WTO successfully in 2000.

③ Other government departments include some of the functional departments from the former state planning commission and the state economic and trade commission.

economic reform, the market system was finally introduced into China's foreign-aided construction projects after 2000, 20 years behind the development of China's domestic circumstances. One of the most influential changes was the implementation of the tender and bid systems in China's foreign-aided construction projects, from design to construction work. Enterprises and institutes involved in the designs and constructions of these overseas projects were selected through specific bid operations, which meant that the use of "one-size-fits-all" approaches had ceased and the assignment mode from China's governments also ended.

The Bidding Committee of Foreign-aided Projects① from MOFCOM was the central management department for the bidding, which is composed of seven members, who were mainly leaders of relevant Chinese governmental departments. For the bidding operation of design works, initially all design institutes in domestic China could participate. According to the scales of the project, the bidding committee selected 10 to 25 enterprises through the qualification examination to participate in the final bidding process, which would be evaluated for the winner by experts② extracted from experts' database of foreign-aided projects' bidding evaluation. In fact, such a mechanism was similar to that of China's domestic market, opened to welcome all Chinese design institutes national-wide to participate. In 2012, The Department of Foreign Assistance③ of MOFCOM issued the list of qualified enterprises and posted the list of qualified enterprises that won bids to implement certain categories of foreign aid projects" online④. Twenty institutions were listed as ones that were qualified to design and manage civil aided construction and industrial aided construction projects. Only the companies on this list were authorized to log in the information management platform of the foreign-aided project⑤ and submited bids for the design and management of China's foreign-aided construction projects. The new mechanism scaled down the numbers of design institutes involved and actually led to relatively fixed winners for specific catego-

① The committee is responsible for projects with cost over RMB 10 million yuan, and the department is responsible for small projects with cost less than RMB 10 million yuan. In fact, most of the projects involved in this study, as well as other China's foreign-aided construction projects, cost over RMB 10 million yuan.

② These experts for bid evaluation were screened nationwide, and the qualified experts were selected into the expert database. Experts for bid evaluation of one certain project were randomly selected in the expert database.

③ In 2018, this department was regrouped as the China International Development Cooperation Agency.

④ This list was announced in the official website of MOFCOM(http://yws.mofcom.gov.cn/article/o/i/201601/20160101234363.shtml).

⑤ This platform was established after 2016 by MOFCOM, which was the only official medium for releasing the bidding information of China's foreign-aided projects, and only the qualified enterprises were authorized the accounts to log in.

ries of aid constructions, such as the BIAD, IPPR and SIADR for stadium projects, which will explained in more detailed in the following parts.

Also, the on-going reform of management mechanisms in China's construction industries was simultaneously affecting its foreign projects. New management modes such as PC① and EPC② started to be applied overseas, firstly in investment projects and later in aided ones (Fig. 4.1). This also generated impacts on designs, if adopting the EPC mode in which the design and construction need to be combined in the bidding operation, and the construction drawing would be completed by the construction enterprises while the affiliated design institute paid full attention to the concept. It seemed that in the new era, the mechanism development of China's foreign-aided projects caught up with the domestic ores.

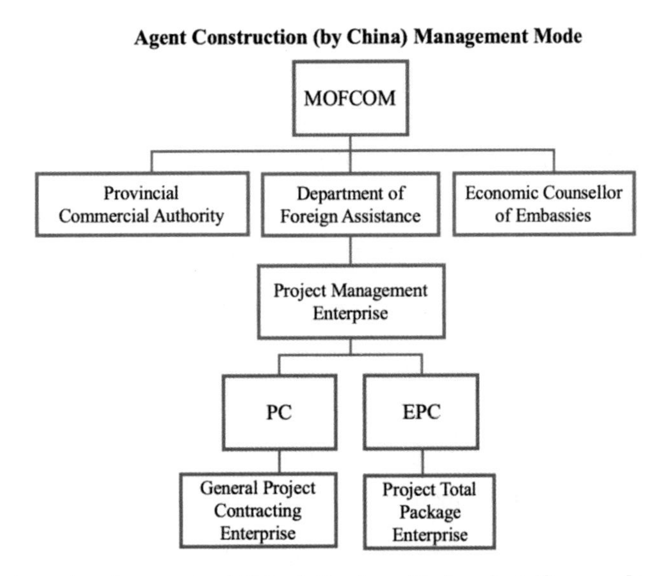

Fig. 4.1 Agent construction mode of China's foreign-aided projects(Source: draw by the author)

① Under the PC (procurement-construction) mode, the project management enterprise undertakes the investigation, design and management work of the project, while the general project contractor enterprise undertakes the construction drawings and the construction works.

② Under the EPC (Engineering — Procurement — construction) mode the project management enterprise undertakes the whole-process management tasks, and the general project contractor enterprise undertakes the investigation, design, construction drawings and general project construction work.

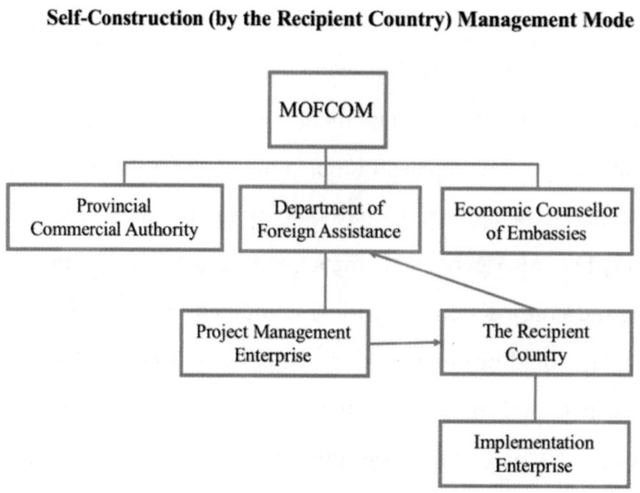

Self-Construction (by the Recipient Country) Management Mode

MOFCOM
Provincial Commercial Authority
Department of Foreign Assistance
Economic Counsellor of Embassies
Project Management Enterprise
The Recipient Country
Implementation Enterprise

Fig. 4.2 Self-construction mode of China's foreign-aided projects (Source: drawn by the author)

The mechanism was in continuous reforming. In the new era the recipient countries participated more in the process, especially for large significant public buildings such as the stadiums (Chang & Xue: 2019a). Moreover, the winning schemes in the bid of China's foreign-aided stadiums were finally selected by the representatives of recipient countries from the alternative schemes chosen by China's domestic experts for the recent significant stadium projects. In the later processes of reviews and constructions, the recipient countries were also gaining more involvement. All these maked the mechanism towards bilateral.

Besides, the self-built mode was put forward after 2015 (MOFCOM, 2015), though almost all foreign-aided construction projects had been completed in agent-construction mode by China before then. In 2018, China International Development Cooperation Agency (CIDCA) was established to take over China's foreign aid affairs, which proposed new suggesting regulations right away (Xinhua Net, 2018). In recent two years, the "localization" mode started to experiment in some China's aid projects of the neighbouring countries that allowed more involvement and responsibilities of the recipient countries, at their requests (Anonymous, 2015). The conclusion chapter will discuss more about this new mode and how it may impact the design, since it may be widely encouraged in China's future overseas aid projects.

4.2 The Blooming of Chinese Stadiums

In the 21st century, China entered an era of rapid development, when Chinese domestic stadiums also face multiple development opportunities promoted by the Beijing Olympic Games.

4.2.1 Blooming by International Sports Events

In 2003, the Chinese Olympic committee won the right to host the 29th Olympic Games, with Beijing as the host city①. This country attached great seriousness to this international sports event to show a good national image to the world. Of the thirty-seven stadiums required, thirty-one were constructed in Beijing (twelve newly constructed, eleven newly expended and eight temporarily constructed). The twelve new stadiums became the main focus and the physical foundation of holding the games. The top priority was the Beijing Olympic Sports Center, including the national stadium, the national natatorium and the national indoor stadium(Fig. 4.2).

In fact, during the first decade of the 21st century, China already hosted several international sporting events, such as the 2001 Universiade in Beijing, the 2007 Asian Winter Games in Changchun and China's national sports events, which required the construction and upgrading of stadiums with higher international standards. Success in the 2008 Olympic Games led to the climax of China's sports career, followed by the 2010 Asian Games in Guangzhou②, the 2011 Universiade in Shenzhen③, the 2014 Youth Olympic Games in Nanjing, 2009 China's National Games in Shandong and 2013 China's National Games in Liaoning, which resulted in continuous construction of high standard stadiums and urban infrastructures. These sports events accelerated the construction of large-scale (over 6,000-seats) indoor stadiums, especially the mega-scale stadiums (10,000-seat or more), to develop from major Chinese cities to other cities in considerable number④. Medium-sized indoor stadiums (3,000-6,000 seat) and small size indoor stadium (less than 3,000) were likewise booming. The construction of mega-scale outdoor stadiums in this period was similarly growing in domestic China. At least 20 mega-scale stadiums had been built inside China.

① Qingdao, Hong Kong, Tianjin, Qinhuangdao, Shenyang and Shanghai were authorized as the supportive cities for the 2008 Olympic Games.

② Twelve new stadiums were designed and constructed for Guangzhou Asian Games, such as the the Asian Games City Gymnasium, the Guangzhou International Sports Performance Center, stadiums of the Zhuhai Sports Complex, etc.

③ 22 new stadiums were designed and constructed for the 2011 Universiade in Shenzhen, such as stadiums of Shenzhen University Town Sports Complex, Shenzhen Bay Sports Complex, Longgang Unversiade Sports Complex and Baoan Stadium.

④ In the past two decades of the new century, approximately 23 mega-scale indoor stadiums had been designed and constructed, such as the new Guangzhou Indoor Stadium (2001) with a 10,000-seat capacity, the indoor stadium of the Xi' an Sports Complex (2020) with 18,000 seats, and China's National Indoor Stadium (2007) for the Beijing Olympics with 20,000 seats.

The capacities of the stadium of Guangdong Olympic Sports Center (2001), China's National Stadium (2008) and the stadium of Hangzhou Olympic Sports Expo City (2018) each exceeded 80,000 seats, and China's National Stadium holds 91,000 seats(Fig. 4.3). More other large-size stadiums, and medium- and small-size stadiums were built in more extensive areas of China.

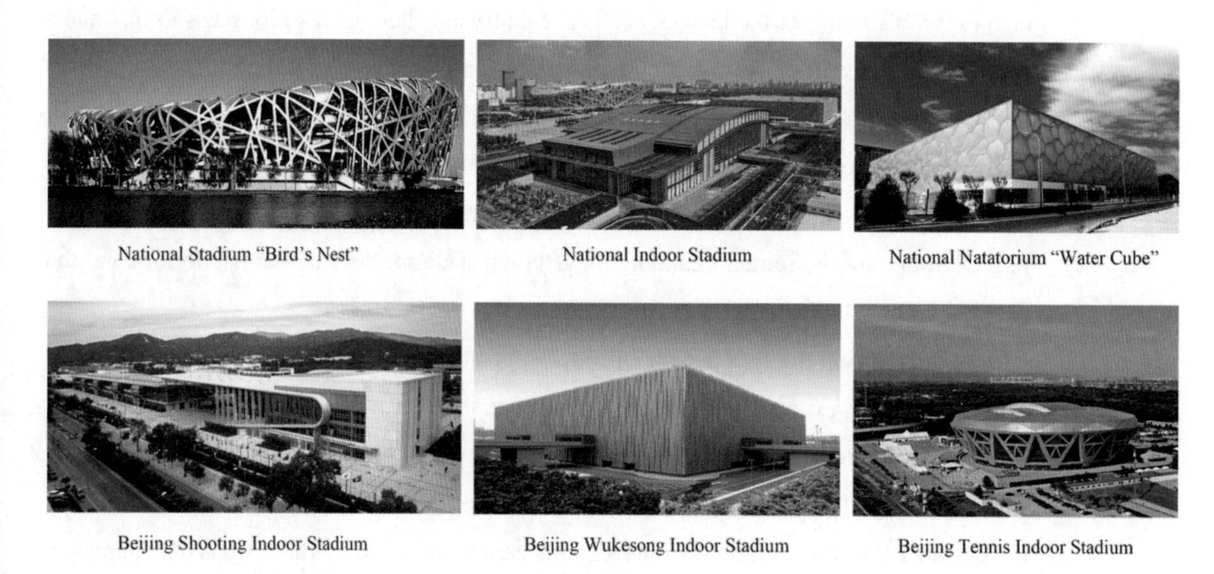

Fig. 4.3 Stadiums constructed in Beijing for the 2008 Olympic Games

Fig. 4.4 Some of China's large stadiums and their capacities of the new era: (a) Guangzhou Indoor Stadium; (b) Indoor Stadium of Xi'an Sports Complex; (c) China's National Indoor Stadium; (d) Stadium of Guangzhou Olympic Sports Centre; (e) China's National Stadium; (f) Stadium of Hangzhou Olympic Sports Expo City.

The newly constructed large indoor stadiums in China mostly adopted a larger site to satisfy the needs of the gymnastic platform in competition. China's new architectural design codes for sports architecture in 2003① expanded the basic size of the playgrounds. Architects started to try larger-sized fields in Chinese domestic indoor stadiums. The requirement of international sports games such as the NBA and FINA had become the main concerns in the design. Besides, with considerations of the post-game operation, Chinese domestic stadiums of the new era stepped towards complexes with commercial, entertainment and cultural attributes. Some stadiums (e.g., the stadium of Changzhou Sports Exhibition Center) adopted the eccentric seating arrangements to give better views for the audience in shows. At the same time, the influence of international sports events on urban development was extended through the construction of sports venues, and stadiums had become driving forces of urban development, and represent the city's modern image to a considerable degree. Expressions and symbolizations in the appearance of stadiums turned to be one of the main considerations in the concept designs.

4.2.2 Designs with Worldwide Perspectives

Since the 21st century, foreign architects have designed about 50 stadium projects in China, blowing the victories of design bid/competitions of numerous Chinese domestic stadium projects by foreign firms, like Nanjing Olympic Sports Complex (HOK, 2005) and the National Stadium "Bird's Nest" (Herzog & De Meuron, 2008). Their concepts and theories in the designs strongly impacted Chinese architects and Chinese sports architecture, expressing the international perspectives, supplanting the Chinese traditional ones. The design practice of Chinese stadiums was completed by these foreign companies in cooperation with Chinese design institutes who were responsible for the construction drawings. The design abilities of Chinese design institutes were improved in the process of such cooperation.

Under the influence of foreign architects, and with the development of the mathematic geometry, computer technology, digital design, construction technology and materials technology, some Chinese architects made breakthroughs compared to the previous Chinese stadiums, the forms of which followed geometrical constraints of the modernized cube, and transformed the forms from classic Euclidean geometry to free forms like subdivision and

---

① China's architectural design codes for sports architecture in 2003 (Design Codes for Sports Architecture JGJ31-2003) has been used till the present without upgrading or revisions yet. The size of the playground was expanded to be 40 m × 70 m and the size of the infield is mostly 40 m × 70 m and above.

topological ones (e.g., the indoor stadium from Guangzhou Asian Games City Sports Complex). At the same time, the concept of "skin" in architecture started to appear in the design of Chinese stadiums. The 2008 Beijing Olympic Games became the manifestation of stadiums① with unique "skin" as the main form features. These Olympic stadiums became the catalyst of "skinned" stadiums in China, as welcomed in the era of consumption with more information and visual impact transmitted in the coverage. In the later practice of Chinese stadiums, the "skin" concept was favoured by both Chinese architects and the clients for both simple geometric forms (e.g., Tianjin University Indoor Stadium and Guangzhou Huadu Indoor Stadium) and complex forms like topological ones (e.g., the Indoor Stadiums of Liuzhou Sports Complex and stadiums of Shenzhen Universiade Sports Complex). (Fig. 4.5)

Fig. 4.5 Examples of China's domestic stadiums with "skin" concepts: (a) Tianjin University Indoor Stadium; (b) Guangzhou Huadu Indoor Stadium; (c) LiuzhouSports Complex; (d) Shenzhen Universiade Sports Complex

4.2.3 Significant Development of Architectural Techniques

The rapid development of structural, material and digital techniques in architecture

① China's National Stadium used 24 trusses of steel frame columns weighing 42,000 tons woven around the stands to form the coverage of a hollow steel grid structure, similar in the form of "Bird's Nest". China's National Natatorium used ETFE material to create about 3,000 irregular blue air pillows, forming the "water cube" skin for the stadium (BIAD, 2008). The Beijing Olympic Basketball Stadium, jointly designed by HOK and BIAD, has a simple cube form with nanoscale coating technic Low-E glass curtain walls decorated by gold perforated aluminum panels to form a unique "skin" looking like the billowing wheat fields.

prepared Chinese stadiums to hold larger scale, more internationalization and more complex forms. Influenced by the concept of the "Green Olympics" proposed for Beijing Olympic Games, more attention to the use of energy-saving materials and technologies.

For the structural aspect, the tensional system was one of the main development directions of large-span spatial structure that favoured by most projects for minimizing the amount of steel usage (Fu, 2008). Later, the mixed structures that combine two or more structural systems with various mechanisms of forces to form a unique and effective structure referred as the "new spatial hybrid structure" (Liu, 2008), were utilized in Chinese stadiums, and can compensate the defects of a single structure and have advantages in adapting to complex forms. The membrane structure has been widely used in the new period, especially in the post-Olympic period, which enabled the heavy stadium to own the sense of "light" and was adopted in numbers of Chinese domestic stadiums. The combination of membrane structure and metal lattice structure was additionally applied to the facades of Chinese stadiums, such as Zhejiang Lishui Indoor Stadium (2017). In addition, the improvement of pneumatic membrane structure in the design of China's National Natatorium (2008) in the adaptability to bad weather (e.g., snow load, sharp objects tears) also expanded the application field of the membrane in other Chinese stadiums[①].

The development of membrane material (PVC, ETFE, PTFE, etc.) introduced more advantages of light weight, light transmission and easy forming of complex surface. Simultaneously, the development of other materials provided more possibilities for the diversified images and functions of stadiums. The metal plate was used in large Chinese stadiums in cambered surface forms given its attributes of light weight, large size and strong plasticity[②]. And the anodic continuous aluminum oxide curtain walls (combined with the Low-E hollow laminated glass curtain walls) formed a smooth and rapidly changing smooth curved shape of the stadium of Guangzhou International Sports and Performance Center. Optimized permeable/semi-permeable materials such as glass were put into more use as the new skin of the stadium to generate transparent feelings, such as in Dongguan Basketball Stadium (2014) and the stadium of Shenzhen Universiade Sports Complex (2018), etc.(Fig. 4.6)

① For instance, Dalian Sports Complex (2012) used three colours (white, blue and transparent) of pneumatic membranes to construct an enclosure gradually closed up in an oblique separated ring, forming the overall fancy image of "whirling" and "dazzling" (Wei, et, al., 2013).

② For example, the triangular metal plate curtain wall in the stadium of Mercedes-Benzes Sports Complex in Shanghai enhanced the futuristic feeling of the "flying saucer" image.

Fig. 4.6 Zhejiang Lishui Indoor Stadium (right); detail views of the Dongguan Basketball Stadium (mlddle);and the stadium of Shenzhen Universiade Sports Complex (left) (Source: http://lishui.house.qq.com/a/20170303/004290.htm?newspc; http://www.gdliangyi.com/h-nd-45-0_31.html?groupId=5&_ngc=-1; http://www.viibrand.com/2013/news/trend/6_2.html)

Digital software enabled the stadium's designs with complicated forms to be modelled, simulated and analyzed with computer tools, such as the 3D modelling software, visualized scripts, parametric design platforms and BIM. Also, the acceptance of Chinese citizens and the local government transformed from traditional regular forms to the appealing of heterogenic forms, as they regard the stadiums as the landmark of the region as well as the business card for the development of the city(Fig. 4.7). The design levels of Chinese institutes have been obviously improved to catch up with the international mainstream under these forces, as illustrated by their recent designs of large stadiums/sports complexes for Chinese cities.

Fig. 4.7 Ningbo Olympic Sports Complex, designed by CCDI, 2020 (Source: https://mp.weixin.qq.com/s/ngRVOho34-XjSJ182oMh0g)

4.2.4 Summary

In the 21st century, the tide of globalization swept over China, and the reform of Chinese society has passed the transitional stage. Under the promotion of significant sports events, especially the Beijing Olympic Games, Chinese domestic stadiums have undergone enormous development. The quantity and scale of Chinese stadiums were increased dramatically, and the functions of stadiums were inclined to be multiple and flexible. The forms were increasingly diversified, including complex organic geometries and "skinned" facades, which were strongly supported by the development of complex structures and technical materials, "digital tools" of design and construction. Chinese stadiums achieved a leap forward from quantity to quality(Table 4.1).

Table 4.1 Summary of the development of China's domestic stadiums in the new era

Time	Scale (Seats)	Architectural Features	Architectural Technique Development	Sports Event
2000s-	Increasing greatly; Indoor stadium: over 10,000 seats; Outdoor stadium: over 60,000 seats	Involvement of foreign architects; Complex geometric forms; The "skin" trend; Multi-functional and international standard	Membrane structure and material development; Digital modeling and simulation	Beijing Olympic Games

4.3 China's Foreign-aided Stadiums: Adjustment and Adaptation through Cooperation and Competition

In the 21st century, China increased its number of foreign aid constructions, among which the stadiums were included. Over 60 stadiums were exported into other countries by China, with about three-fourths outdoor stadiums and the rest indoor. The regions that received the stadium gifts expanded from Asia and Africa to more continents, such as Oceania and Latin Africa, especially before the 2010s. After the 2010s, the stadiums still mainly focused in Africa and Asia. Generally, 60% (36 stadiums) were located in Africa while over 10% (7 stadiums) were in Asia, echoing of the famous quote from IOC's former president Samaranch about China's foreign-aided stadium in Africa.

Since China's aid diplomacy transformed from grants to cooperation with various financial supporting modes, scales and standards of these stadiums also increased significantly compared with the previous periods. The number of large stadiums with over 30,000-seat capacities reached fifteen, including eight over 50,000-seat. Profound improvement of China's foreign-aided stadiums was achieved to meet a high standard of modern sports buildings in the new era to hold international games. However, the indoor stadium's capacity stayed similarly, for over half of them have no more than 5,000-seat capacities. (Table 4.2)

4.3.1 Transition, Breakthrough and Adaptation under New Circumstances

After 2000, the opening of the market system in China's economic reform spread finally to the aid construction mechanism. China's foreign-aided stadium projects stepped into the tender and bid mode. The new bid system provided chances for more Chinese design institutes to be involved in these foreign-aided stadium projects in the first decade of the 21st century, even including private firms (e.g. CCDI) included. Architectural designs of these stadiums benefited from various participants with diversified schemes. The designers of these stadiums transformed from chief architects of the institute to architect teams consisting of younger technicians of the specialized departments that had been established for oversea projects, and the working mode tended to be parallel with that of commercial projects. After the issue of qualified enterprises for the bid of China's foreign-aided construction project by MOFCOM in 2012, some design institutes stepped away from the stadium projects, with only several left as resident ones for the designs of foreign-aided stadiums.

Another transition lies in the working mode of architects that turned out to be more flexible, with multi-cooperation, especially with overseas companies. Although in the 1990s Chinese architects had experienced the cooperation with foreign design firms in aid stadium projects, all the cooperation projects were the cricket stadium, as regulated by the MOFCOM that only their qualified enterprises could design the stadiums, which did not happen in other types of China's foreign-aided stadiums. However, in the 21st century, design cooperation started to commonly appear in normal stadium project, such as the My Dinh National Stadium① in Hanoi and Mahinda Rajapaksa International Cricket Stadium② in Jamaica (Fig. 4.8). When some recipient countries desired more modernized stadiums to represent their achievement in the pursuit of modernization, they might prefer to use the designs of their local architects or foreign firms from the developed countries, but with financial and other technical support from China. Therefore, Chinese architects and Chinese design institutes' roles in the cooperation were the detailed and technical drawing development based

① It was designed by Hanoi International Group with SXDAD.

② It was designed by SWA with SXDAD.

Table 4.2 Information of China's foreign-aided stadiums after 2000

No.	Year	Continent	Country	Project	Design Institution	Capacity (seats)	Photo
Outdoor Stadium							
1	2000	Latin America	Saint Lucia	Beausejour Stadium (Darren Sammy Cricket Ground)	—	15,000	
2	2000-2002	Asia	Vietnam	My Dinh National Stadium	Shanghai Xian Dai Architectural Design (Group) Co. Ltd. (SXDAD) With Hanoi International Group	40,000	
3	2002	Africa	Central African Republic	Barthelemy Boganda Stadium (National Stadium)	—	20,000	
4	2003	Africa	Tanzania	Benjamin Mkapa Stadium (Tanzania National Stadium)	BIAD	60,000	
5-7	2004	Africa	Djibouti	Three small stadiums	—	—	—
8	2004	Africa	Equatorial Guinea	Estadio de Bata (Bata Stadium)	—	35,700	
9	2004	Africa	Guinea	Stade de l'Unité (National Stadium)	CSADI	50,000	
10	2005	Latin America	Bahamas	Thomas A. Robinson National Stadium	—	15,000	
11	2005	Latin America	Jamaica	Sabina Park (Cricket Ground)	HOK & SXDAD	15,600	
12	2005	Latin America	Antigua & Barbuda	Sir Vivian Richards Stadium	HOK & Hunan Architectural Design Institute Ltd.	10,000	

cortinued

No.	Year	Continent	Country	Project	Design Institution	Capacity (seats)	Photo
13	2005	Latin America	Dominica	Stadium of Windsor Park(National Cricket Stadium)	Wuhan Architectural Design Institute (renamed as GIADR①)	12,000	
14	2006	Africa	Equatorial Guinea	Estadio de Malabo (Malabo Stadium)	—	15,250	
15-17	2006	Africa	Mali	Three small stadiums	—	—	—
18	2007	Africa	Zambia	Levy Mwanawasa Sports Stadium	BIAD	49,800	
19	2007	Latin America	Grenada	National Stadium	China International Engineering Design & Consult Co., Ltd. (CIEDC)	10,000	
20	2008	Asia	Laos	Laos National Stadium	CCDI	2,5000	
21	2008	Africa	Mozambique	Estádio Nacional do Zimpeto (new National Stadium)	Wuhan Architectural Design Institute (GIADR)	—	
22	2009	Oceania	Papua New Guinea	Wewak Sports Stadium②	Hunan Architectural Design Institute Ltd.	—	
23	2009	Africa	Gabon	Stade d'Angondjé (National Stadium)	*Shanghai Construction Group*	40,000	
24	2009	Latin America	Costa Rica	Estadio Nacional de Costa Rica (National Stadium)	CSADI	35,000	

① General Institute of Architectural Design and Research Co., Ltd.

② The stadium was built at a cost of 19 million Kina, the combined contribution of 12 million Kina in funds from Chinese government and 7 million Kina contributed by Papua New Guinea. (Gare C. "Graffiti Goes on Stadium", Mamose Post)

cortinued

No.	Year	Continent	Country	Project	Design Institution	Capacity (seats)	Photo
25	2011	Africa	Zambia	Gabon Disaster Heroes National Stadium	BIAD	50,000	
26	2011	Africa	Congo	Stade de Djambala	—	—	
27	2011	Africa	Congo	Ewo Stadium			
28	2012	Africa	Senegal	Stade Municipal de Mbour	—	5,000	
29	2012	Africa	Senegal	Stade Mame Massene Sene De Fatick	—	—	—
30	2012	Africa	Cape Verde	Stadium of Cape Verde	CIEDC	10,000	
31	2012	Africa	Congo	Brazzaville Stadium (in Brazzaville Sports Complex)	CCDI	60,000	
32	2012	Africa	Ghana	Cape Coast Sports Stadium	IPPR	15,000	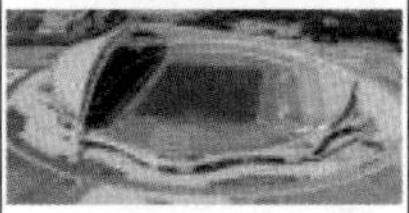
33	2013	Africa	Malawi	Bingu National Stadium	BIAD	42,900	
34	2014	Africa	Sierra Leone	Bo Stadium	—	4,000	
35	2014-2017	Africa	Cote D'lvoire	Olympic Stadium in Abidjan	BIAD	60,000	

cortinued

No.	Year	Continent	Country	Project	Design Institution	Capacity (seats)	Photo
36	2015	Africa	Comoros	Moroni Stadium	IPPR	10,000	
37	2015	Asia	Cambodia	National Stadium	IPPR	60,000	
38	2016	Africa	Gabon	Stade de Port-Gentil	China State Construction Engineering	20,000	
39	2016	Africa	Gabon	Stade d'Oyem	SIADR	20,031	
40	2017	Africa	Ethiopia	Addis Ababa National Stadium	MH Engineering PLC①	60,000	
41	2017	Africa	Senegal	National Wrestling Stadium	IPPR	20,000	
42	2018	Africa	Mauritius	Stadium of Centre Culturel et Sportif (multi-purpose sports complex at Saint Pierre)	BIAD	15,000	
43	2019	Africa	Chad	Stade de N' Djamena (Stadium of Djamena)	SIADR	30,000	
44	2019	Europe	Republic of Belarus	National Football Stadium	BIAD	33,000	
Indoor Stadium							

① LAVA won the design competition of the stadium, but the final design was changed and completed by MH Engineering PLC, a famous architectural design company of Ethiopia. This stadium was constructed by China State Construction Overseas Development Co. Ltd.

cortinued

No.	Year	Continent	Country	Project	Design Institution	Capacity (seats)	Photo
1	2002	Oceania	Fiji	Laucala Bay Gymnasium	China Huashi Group Co., Ltd.	3,200	
2-4	2004	Africa	Morocco	Three small natatoriums (in Fiss, Kenitra)	CADRG	—	
5	2004	Oceania	Samoa	Samoa Aquatic Center	Realway Engineering Consulting Group Co., Ltd.	—	
6	2007	Africa	Cameroon	Sports Palace of Yaoundé	—	5,400	
7	2007	Asia	Mongolia	Buyant Ukhaa Sport Palace (National Indoor Stadium)	CIEDC	5,045	
8	2008	Asia	Laos	One Natatorium and two Indoor Stadiums of the National Sports Park	CCDI	—	
9	2010	Asia	Sri Lanka	Mahinda Rajapaksa International Cricket Stadium	SWA① with SXDAD	35,000	
10	2012	Africa	Congo	Brazzaville Indoor Stadium (in Brazzaville Sports Complex)	CCDI	10,000	
11	2012	Africa	Congo	Brazzaville Natatorium (in Brazzaville Sports Complex)	CCDI	2,000	
12	2017	Africa	Tunisia	Natatorium (Ben Arous Youth Sports and Culture Center)	SIADR	—	

① This stadium was constructed by China Communications Construction Co., Lt.

cortinued

No.	Year	Continent	Country	Project	Design Institution	Capacity (seats)	Photo
13	2017	Africa	Gabon	Libreville Indoor Stadium (for handball games)	—	5,477	
14	2017	Africa	Mauritius	Natatorium of Centre Culturel et Sportif	BIAD	1,100	—
15	2018	Africa	Algeria	Youth Sports Center	IPPR		
16	2019	Europe	Republic of Belarus	International Standard Natatorium	BIAD	6,000	

on the conceptual designs from the other party. Ever since the announcement of the Beijing 2008 Summer Olympics Games, China has been conducting intensive construction of sports venues consistent with the high international standards not only in its capital but also in other major cities, which in turn has improved the design and construction level of sports buildings constructed in the aid program. Chinese architects and design institutes got more practice of designing high standard stadiums, and coincidently returned to the court of designing foreign-aided stadiums independently after the 2010s.

Fig. 4.8 My Dinh National Stadium (left); Mahinda Rajapaksa International Cricket Stadium (middle); Addis Ababa National Stadium[①] (right) (Source: left, the photo was taken by the author in 2019; middle, provided by SIADR; right, https://architizer.com/projects/addis-ababa-national-stadium-sports-village/)

Under such transitions, China's foreign-aided stadiums in the new era tend to be more diversified with higher standards, larger scales, and more powerful influence. Some of the ever-best China's foreign-aided stadiums were designed and constructed through Chinese

① It was designed by MH Engineering PLC.

architects and engineers. Over ten stadiums used full-coverage roofs (Fig. 4.9), and many had tried the "skin" language in architectural expression, which required higher cost and more complex structures and materials. In the first years of the new century, most of China's foreign-aided stadiums still set up basic functions for holding the sports events. In the recent decade, the functions, as required by the recipient countries, were also diversified, combining with more office and commercial space. In addition, the new types of multi-functional sports centers appeared such as the China-aided Youth Sports Center in Algeria and Ben Arous Youth Sports and Culture Center in Tunisia. There are also new specialized stadiums appearing in China's foreign-aided stadiums, such as the wrestling stadium and cricket stadium, which were seldom designed and constructed in domestic China, providing Chinese architects with opportunities to experiment chance and preparing Chinese stadiums to be more diversified and expertized.

Fig. 4.9 Examples of China's foreign-aided stadium with full coverage roof of the 21st century

For both outdoor stadiums and indoor stadiums, the general forms were relatively regular compared with China's contemporary domestic stadiums, many of which experimented complex geometries with the help of parametric designs. This might be partly because the economic limitation that remained sometimes, and more, as the author believed, because what the recipient country preferred maybe more conservative about the principle of delight and modernity. Even for the design by foreign firms, for instance, the final scheme of China-aided Addis Ababa National Stadium of Ethiopia was changed to be the regular-form scheme of MH Engineering PLC from the original winning scheme of LAVA. (Fig. 4.10)

Fig. 4.10 LAVA's winning scheme of Addis Ababa National Stadium of Ethiopia (left); the final scheme of Addis Ababa National Stadium of Ethiopia by MH Engineering PLC (right) (Source: https://www.l-a-v-a.net/projects/addis-ababa-national-stadium-and-sports-village/; http://stadiumdb.com/designs/eth/addis_ababa_national_stadium)

Recently, many recipient countries expectd these buildings designed and constructed by China to be the symbolization of the modern level of the city/country and to be qualified for the application of significant international sports event and other national events. The bidding mechanism forced Chinese architects to satisfy the recipient country's preference, especially when it became the final decision-maker during the reforming of China's foreign aid. Chinese architects started to introduce more regional design approaches, with attention to the local culture and climate, in their understandings in the exploration of modern design languages. Compared with the previous two periods, such regional concerns in the architectural design of stadiums tend to break the economic limitation and conservative routines. More large-scale, multi-dimensional and high-profile expressions of regional aspects were integrated in the outdoor stadiums, such as the Cambodian new national stadium and other cases illustrated in the case study section of the present chapter. And these efforts gradually transformed to more adaptive ones that appealed the recipient countries with the recipient countries' participating more in the design process.

Another remarkable adaptation approach lies in the using of Chinese standards. According to previous regulations by MOFCOM, the design, management and construction of all foreign aid projects must comply with Chinese standards and Chinese building codes. This requirement generated convenience for Chinese technicians' work and the exportation of Chinese products and labours abroad, which also helped to export the Chinese standards abroad for future development. This reminded Chinese architects the challenges of adapting

Chinese standard to local circumstances, such as habits, specifications and criteria when designing these mega-structures to make the designs more satisfactory. Such adaptation was also encouraged in the "Eight Guiding Principles for Design"①, proposed in the MOFCOM manual for China's construction aid projects. Among these guiding principles, "standard application" was listed first. For stadium projects in less-developed areas, where formal building standards barely exited, Chinese architects need to consider how to better integrate Chinese standards with local habits and customs. Some post-colonial countries still adhered to standards, most of which were European or American standards, that had been imposed by these countries' former suzerains. Chinese architects need to pay special attention to dealing with the conflict between Chinese standard and others in the design process, such as in the designs of China-aided national stadium of Costa Rica by CSADI (Fig. 4.11).

Fig. 4.11 China-aided national stadium of Costa Rica (Source: provided by SIPPR)

4.3.2 Case Studies

Since the number of China's foreign-aided stadiums has increased greatly with its development of higher qualities and diversified designs, more cases were selected in the present section for comprehensive understandings. The national stadium of Tanzania was designed in the early years of the 2000s and also was the first case in which Chinese architects cooperated with a foreign design company. While the design work of China's foreign-aided stadium in Ivory Coast lasted from the later 2000s till the early 2010s, and the last two cases were designed in recent years. These cases could illustrate the most recent development of architectural features of China's foreign-aided stadiums during the years. More importantly, these cases were designed by different architectural design institutes and have revealed dif-

① "Eight Guiding Principles for Design" includes "standard application", "overall planning", "investment matching", "function priority", "technological innovation", "environmental protection", "convenient maintenance" and "sustainable development".

ferent aspects in the design process. Therefore, particular case studies were conducted on these China's foreign-aided stadiums.

4.3.2.1 Adapting Standards and Techniques: National Stadium of Tanzania

As it established diplomatic relations with China in 1964[①], Tanzania received a large amount of construction aid from China such as factories and collieries[②], among which was China's first large-scale aid project, the Tanzania-Zambia railway of 1,860 km long in 1975. For this newly independent country from British rule, sports played a crucial role in its nation building process, and the stadium became a significant venue where national events were held such as the National Liberation Day.

The Tanzania government planned for the construction of a new national stadium in the early 2000s, but the plan stalled due to financial insufficiency and the forbidden pressure from World Bank and IMF. Soon China stepped in and offered to partially finance the stadium.

China's reform of the financial aid mechanism supported the profound improvement of China's foreign-aided stadiums to meet a higher standard of modern sports buildings in the new era. Unlike that project from the early stage of China's aid program, the Tanzania National Stadium project was jointly financed by the Chinese (approximate US$ 35 million) and Tanzania governments (US$ 23 million), for a total cost of over US$ 58 million. The cooperation of finance brought the cooperation in architectural designs. This stadium was one of the first China's foreign-aided stadiums that were designed in cooperation with foreign enterprises in the new century. The site planning was completed by BKS Group, and the schematic design of the main stadium was designed by WAS Architects, both of whichwere companies from South Africa appointed by the Tanzania government of the design work (Fig. 4.12). The tender and bid system was not applied in the design yet, but was in construction of China's domestic enterprises for undertaking the missions. When the Beijing Construction Engineering Group (BCEG) won the construction bid, Beijing Institute of Architectural Design (BIAD) was the key partner in developing the conceptual design and completing other technique support and drawings, as stated in the contract. Such cooperation enabled this stadium to be more modernized and internationalized, meeting both the International Association of Athletics Federations (IAAF) and Fédération Internationale de

① China established diplomatic relations with Tanganyika on 9 December 1961 and with Zanzibar on 11 December 1963. After their uniting, China continued the diplomatic relations with Tanzania.

② The main projects include the Tanzania-Zambia railway, the Friendship textile factory, the Mbarali farm, the Gewerna coal mine and the Mhongda sugar factory.

Football Association (FIFA) standards, as proposed by Tanzania in the contract with China.

Fig. 4.12 Original designs of DSM sports complex

(Source: https://ilovetanzania.blogspot.com/2010_06_01_archive.html)

Designed by two South African companies, the initial scheme followed South African standards and adopted the usage habits of the Tanzania local. However, the detailed and technical designs (including structures and materials) by BIAD needed to use Chinese standard as most China's foreign-aided construction projects did, for the construction to be led by BCEG in corporations with Tanzanian contractors. This required Chinese architects to adapt Chinese standards into the local ones and that used in the initial design. Although it was not the original creation completed by Chinese architects, how Chinese architects tried to adapt Chinese standards to the local designs through their developing designs is interesting and worthy of discussions.

Located in Dar es Salaam next to Uhuru Stadium, Tanzania's previous national stadium, the new national stadium, also known as Benjamin Mkapa Stadium, has a 60,000-seat capacity and 69,050 m^2 gross floor areas. With the considerations of Tanzania's love for football games and the low use frequency of track and field competitions, the shape of the stands was designed to hold two straight east-west edges and two semicircles to minimize the horizontal distance between the audience and the site for football games. This is quite different from the four circle-centre shape as commonly used in China's domestic. The general layouts of the floors, the roof forms and the surrounding facades also followed the shape of stands. The under space of the upper-layer stand was kept opened without the arrangement of additional rooms, unlike China's domestic large-scale stadiums in which the under-stand space was normally fully utilized by numbers of functional rooms such as hotels and offices. Only the lower level under-stand space was filled by some auxiliary rooms. The in-depth designs formed the stadium to be pure and concise (Jiang, 2007).

Another significant difference existed in the barrier-free designs. As over 10% of the local population was disabled, the conceptual design placed two large cross slopes on the north and south sides following the semicircle shape, and one centralized slope at the main entrance to connect the ground floor and first floor. The idea was well conveyed in the in-depth design by Chinese architects and was even improved to better serve the capacity of traffic to the stadium by replacing the single slope of the main entrance with two separated ones alongside (Fig. 4.13). The adjustment was also in coincidence with the site designs developed by the Chinese institute with symmetric axial squares and parking areas. Besides, more quantities of seats for disabled people were arranged in this overseas stadium, which was of a higher standard than China's domestic barrier-free design standard.

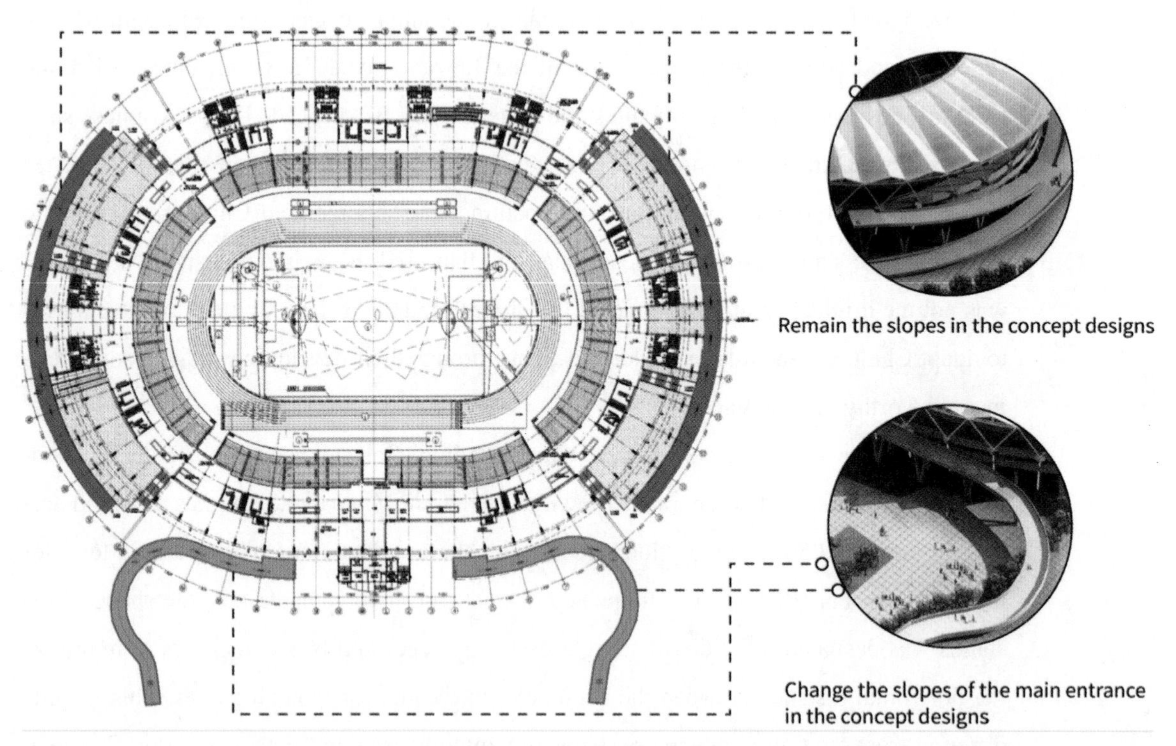

Fig. 4.13 Improvement of the designs by Chinese architects in China-aided national stadium of Tanzania (Source: drawn by the author based on pictures from National Stadium of Tanzania. Archicreation, 91(1):50-55. Jiang, 2007)

In addition, standards were not the only challenges that faced Chinese technicians. To portray a modernized stadium with advanced technologies, WAS Architects' initial designs used the membrane roof fully covering the two-layer stands, together with the mainstream development of stadiums worldwide. To achieve the modern, international and high-tech image, more advanced structure and technologies were utilized in the in-depth design by BIAD. Despite the traditional reinforced concrete structural system as frequently used in former China-aided stadiums such as the Moi Sports Centre, the spatial pipe truss for the main perpendicular and roof structure, and the cable-membrane tension structure were used in the stadium. Chinese architects introduced more V-shape supporting columns than indicated in its initial designs. However, the repeating curved membrane roof was replaced by the folding triangular plane-shape membrane roof with V-shape section, which simplified the membrane structure and material requirements (Fig. 4.14, Fig. 4.15). Nevertheless, the structural teams of BIAD still tried hard on the structural designs (Meng & Yan, 2011; Anonymous, 2006), as the membrane structure was not commonly used in Chinese domestic stadiums at the beginning of the 21 century. New structure reduced the weight to generate more lightsome appearance. Also, the roof was constructed with an advanced ETFE material that had heat-resistant abilities (with solar reflectance above 70%). To improve the rainfall shortage in Dar es Salaam, a special rainwater recycling system was set up (Jiang, 2007).

Such simplification and energy-efficiency approach seem to be the inheritance of the design routines from the 1980s—1990s period, when Chinese architects used economic methods to achieve the general effects. However, the higher requirements and international cooperation of the foreign-aided stadiums forced them to pursuit better design. The ETFE material and membrane structure were first used in China-designed/constructed stadiums in advance of Chinese domestic stadiums (Fig. 4.16), when several years later Beijing's "Water Cube" national natatorium shone its light using the same advance material and structure. The author believed that the experiment of the new material and structure in this China-aided stadium might provided a practical reference for the spread using of the membrane in Chinese domestic stadiums after 2008, and also prepared BIAD for its future wining of the design of more other China's foreign-aided stadiums such as the Cote d'Ivoire stadium.

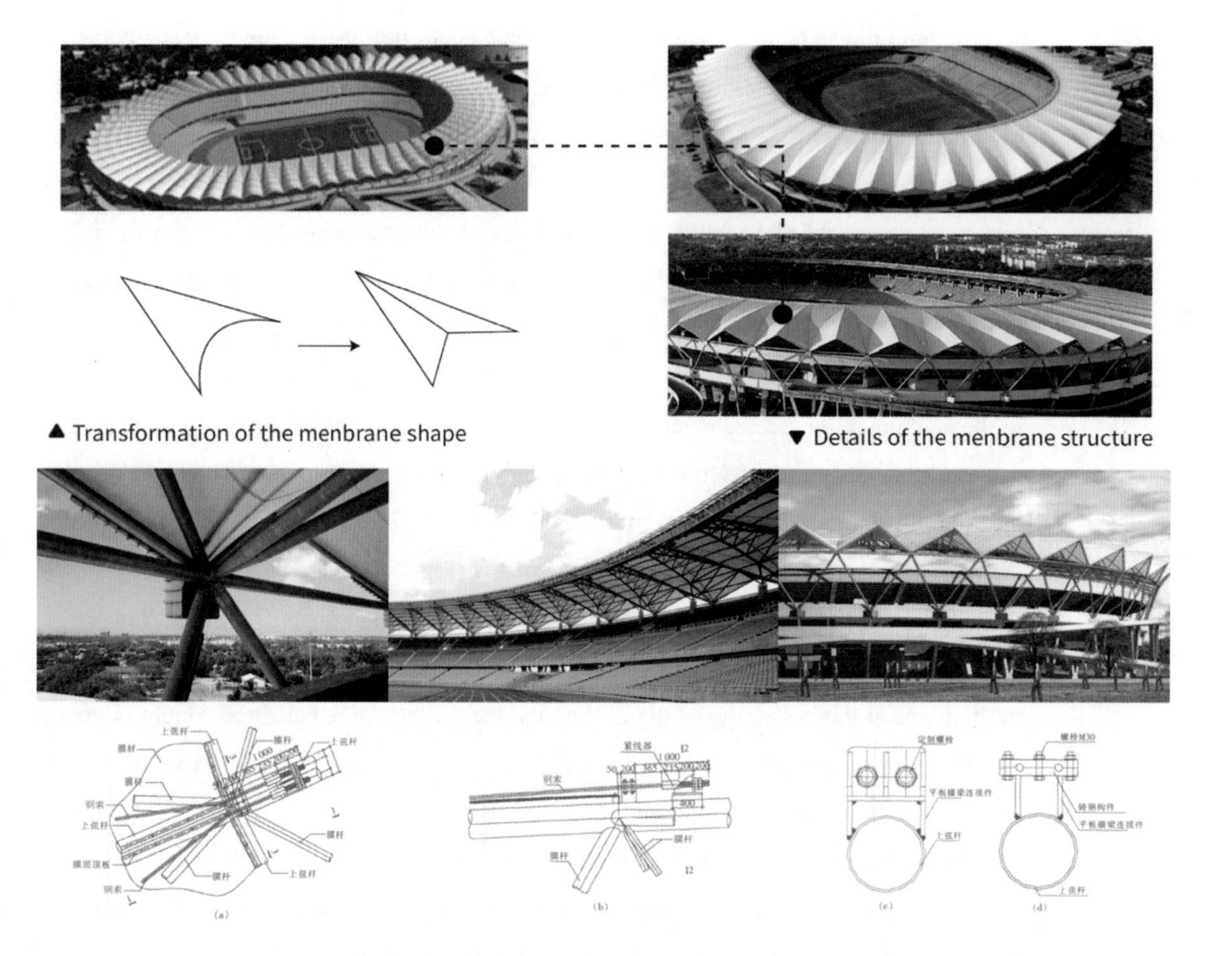

Fig. 4.14 Membrane structure analysis of China-aided national stadium of Tanzania (Source: drawn by the author based on pictures of National Stadium of Tanzania. Jiang, 2007;Meng and Yan, 2011)

Fig. 4.15 The structures with V-shape elements of China-aided national stadium of Tanzania. (Source: http://stadiony.net/stadiony/tan/benjamin_mkapa_stadium)

Fig. 4.16 The bird view (left) and inside view (right) of the Tanzania National Stadium (Source: https://www.stadiumguide.com/tanzania-national-stadium/)

Top-level Chinese and Tanzanian politicians and local officials paid special attention to its construction. Chinese government bailed this stadium as another sign of the historically strong Sino—Tanzanian relationship besides the Tanzania—Zambia Railway project. Chinese Premier Wen Jiabao took time to visit the construction site of the new national stadium during his visit to Tanzania in 2006 (Cooperation 2007; Philamon 2007). Tanzania National Stadium hosted its first events in 2007 after construction completed, and was officially inaugurated in 2009. The stadium is at present the home of Tanzanian top sides Simba and Young Africans, and has replaced the Uhuru Stadium as the national team home ground. Behind the officials' rhetoric of friendship and cooperation, this stadium revealed the mode of China's aid in the 21st century transferring to market-oriented, which benefited both sides for Chinese exportation of designs, labours and materials, and Tanzania's experiences in techniques and constructions through involvement.

4.3.2.2 Understanding of the Local and the Design between Bio-influence: China-aided Stadium in Ivory Coast

Ivory Coast established its diplomatic relationship with China in 1983. It has received construction aid from China ever since, such as the Senator's Home, the Conference Hall of Ministry of Foreign Affairs, hospitals and schools, etc. In recent years, Ivory Coast has been one of Africa's most dynamic economics[①] and China has become its largest financing country and the third greatest trading partner.

In 2014, Ivory Coast aimed to hold the 2021 Cup of African Nations (CAN), and three stadiums were planned to be constructed in three different cities: its capital Yamoussoukro, its largest harbour city Abidjan and its second-largest harbour city San Pedro. The Stade de Yamoussoukro (Stadium of Yamoussoukro, 2018) was a 20,000-seat football stadium and

① Ivory Coast's GDP(Gross Domestic Product) growth in 2019 is 9%.

was designed by SCAU, the French design firm well renowned for its design of the Stade de France (Stadium of France) in Paris and the Olympic Stadium Ataturk in Istanbul (Fig. 4.17). The capital's new stadium cost 60 million dollars and was financed by the Ivory Coast government. The 20,000-seat Stade de San Pedro (stadium of San Pedro) was also financed through a commercial loan from China's banks① and constructed by China Civil Engineering Construction Corporation under the commercial contract. The most costly and largest-scale stadiums of the three, the Olympic Stadium of Ebimpe, as usual, was aided by China through low-interest loans, both designed and constructed by Chinese enterprises.

Fig. 4.17　Stade de San Pedro (left); and Stade de Yamoussoukro (right)
(Source: https://www.dezeen.com/2018/04/03/sloping-ring-shaped-roof-will-cover-scaus-stadium-for-africa-nations-cup-2021/; https://en.wikipedia.org/wiki/Stade_de_San_Pédro)

The new China-aided stadium was located in the northern entry gate of Abidjan, Ivory Coast's informal capital, in suburban areas Ebimpe and Anyama. Abidjan currently had just one international-standard ground, in the 35,000 capacity Houphouet-Boigny stadium built before its independence in 1960, which only met most basic structural criteria and was not in a good state. In addition, the new stadium was a part of the Olympic Village Ebimpe, a multi-use governmental level project with a total land of 287 hectares, which was expected to become the centerpiece of the 2021 CAN and was funded through private investments and public procurement, partly supported by a Dutch support body, the International Service Group (ISG)②. A highway would also be constructed to link the area to Abidjan and the Ghana border. (Fig. 4. 18)

① The stadium by commercial loans from Chinese banks was excluded from China's foreign-aided stadium in the research, for it was not issued from MOFCOM without aid purposes.

② https://www.thisdaylive.com/index.php/2019/12/31/caf-2023-ivory-coast-signs-2-5bn-euros-olympic-village-project-deal/.

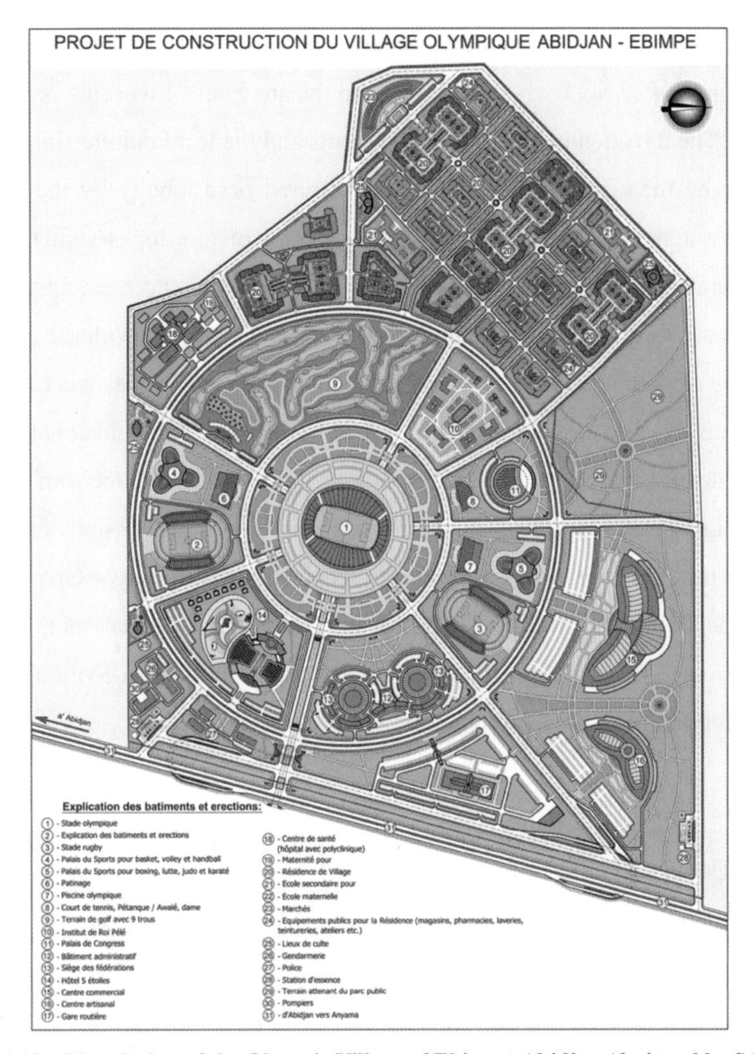

Fig. 4.18 Plan design of the Olympic Village of Ebimpe, Abidjan (designed by SABA)

(Source: https://savm-saba.com/olympic-village-ebimpe-abidjan/)

This China-aided Olympic stadium held a 60,000-seat capacity and RMB 0.75 billion-yuan cost, covering 20 hectares. It required to meet the standards of holding international high-level football, athletics and rugby tournaments games and to be one of the largest and most modernized stadiums in Africa after construction. Both the design and construction were put into biding in domestic China by evaluations to determine the enterprises for the missions, as the results of the newly developed mechanism of China's foreign-aided construction project of the new era.

BIAD won the bid for designing the new national stadium project in 2015, with the concept of "African drum" (Fig. 4.19). As introduced by its chief architect Liu Miao in the interview, the idea of a "drum" suddenly occurred to him when he was watching one football game at night after work while relaxing from the anxieties of working on the design of the stadium[1]. The devotion of Africans to the sports and the local culture filled the mind of the designer, and the appropriate concept was inspired occasionally by the sound of the game he was watching. The main image of this mega-structure looks similar to the drum with local ethnic characteristics. The emphasizing of symbolic and metaphorical forms attracted the attentions of the bid evaluation experts again. Compared with the two other new stadiums to be constructed in Ivory Coast designed by foreign firms, the China-designed one holds relatively obvious characteristics shared in the new China-aided national stadium of Tanzania, identically designed by BIAD with using of the membrane roof in full circles and the waving-feeling supporting structural elements. The initial designs of this stadium also illustrate the influence Chinese architects received from Beijing's Olympic stadiums (Herzog and de Meuron's "Bird Nest" and "Water Cube", for instance) and their favour of the "skin" coverage in the designs of stadiums from domestic China. Symbolization in the forms of stadium "skins" was exported into China's overseas projects.

Fig. 4.19 The initial winning design by BIAD of China-aided stadium in Ivory Coast (Source: http://www.biad.com.cn/newspost.php?id=22)

When the winning scheme was presented to the recipient country, it faced with complete revision. The "African drum" concept was totally replaced by a new one dubbed "Arc de Triomphe" (Triumphal Arch). As one of the football-world powers and the winner of CAN, this country's enthusiasm in football games represents its nation and spirits to a great extent. Chinese architects had noticed the significance of football, but they expressed the

① It was stated by Liu Miao in the interview by the author in 5 August 2018.

feelings in a mild method, which, however, the recipient country would like to portray in a contrary or more obvious way. The supporting structural elements of the facades were changed into tensioned lines upward from the base to imitate the shapes abstractly when people stretched their bodies with shoulders on shoulders as a metaphor of power and unity. The curved convergences of two lines on top formed the Triumphal Arch in deformation shapes, which enabled the general appearance of the facades to echo the winning cup of CAN. Decorations on the facade were designed with orange interior walls, white rods and green plinth, which matched the colours of the national flag and highlights the national image of the recipient country, conforming to the identity of being a new nation (Fig. 4.20).

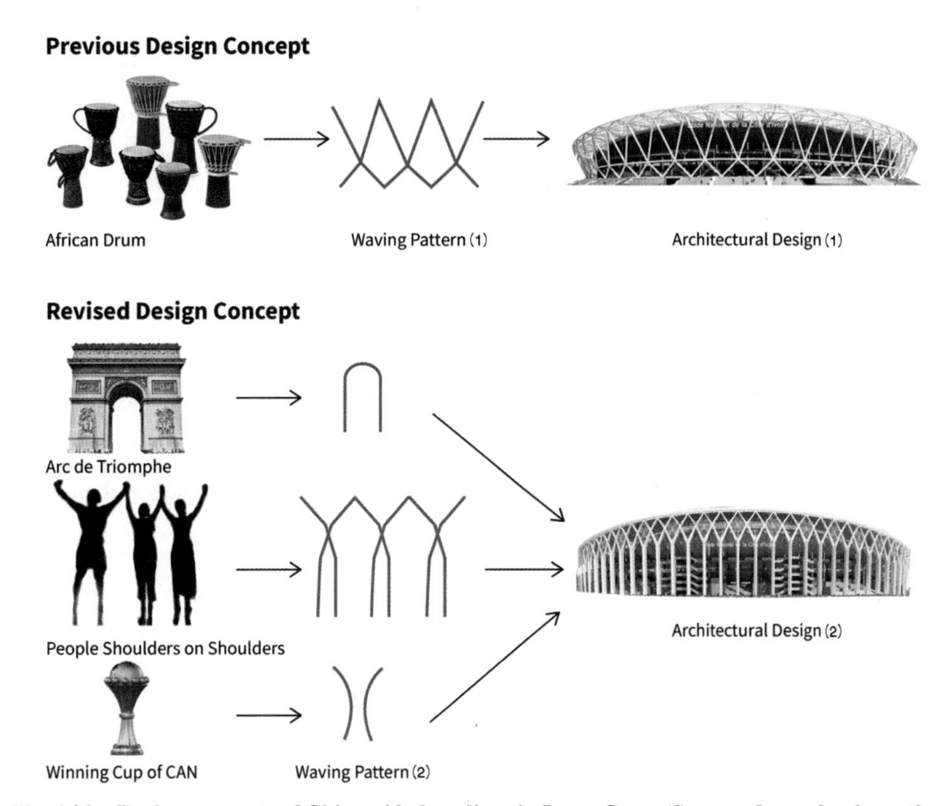

Fig. 4.20 Design concepts of China-aided stadium in Ivory Coast (Source: drawn by the author)

This case illustrates that the recipient country started to care about what the gift from China would look like and involve itself in the design of the donated stadium. Although China's experts made the final decision in the bidding evaluation procedure in domestic China, the recipient country's opinions may also influence the final results greatly through the revision process. Such opinions sometimes interfered with the design and even totally changed it (Fig. 4.21). The revisions with strong opinions from the recipient country expanded the working time of the design of this stadium from 2014 to 2017. This also reflected the problems that happened when the new tender and bid mechanism in China's foreign-aided construction projects were under the absolute control of Chinese experts and governmental departments that the final winning designs by Chinese architects might not satisfy the recipient side. To improve the design efficiency and solve the problems of misunderstanding, a new adjustment was applied soon after, as explained in the following cases.

It can be revealed from this stadium that standard improvement was made for China's foreign-aided stadiums after the 2010s. The China-aided stadium in Ivory Coast had three-layer stands① and full coverage membrane roof with a higher cost, more multiple functions and better design standards for international games. Funded through an EIBC (Export-Import Bank of China) loan, the construction was completed by Beijing Constructor Group by the end of 2019 after 34 months from the ground-breaking, well ahead of the 2021 CAN, which was staged in Ivory Coast(Fig. 4.21). The acceptance of the stadium was also conducted by Chinese enterprise (Zhengzhou Zhongxing Project Supervision Co., Ltd) before it was delivered to the Ivory Coast government. The author regards this stadium as a turning point that the designs of China's foreign-aided stadiums transformed from economic ones to high-standard ones with regional expressions.

4.3.2.3 High-profile Ingratiation: National Wrestling Stadium of Senegal and New National Stadium of Cambodia

In recent years, the development of the tender and bid mechanism in China's foreign-aided construction projects has given recipient countries more discourse rights in the process based on their presence as one of the final decision-makers to choose the design scheme. Most of these countries, who succeeded their independence from long-time colonization in the 20th century or even in the 21st century, tend to express a strong consciousness of nationality and culture in these stadiums. Chinese architects and design institutes tended to accommodate with such desires for winning the bid through more obvious expression and high-profile ingratiation in architectural designs. This can be well illustrated by the two following cases designed by IPPR: the National Wrestling Stadium of Senegal and the new National Stadium of Cambodia.

---

① According to the author's statistics, this stadium is the second three-layer stand stadium of China's foreign-aided stadiums. The first is the China-aided Brazzaville Stadium in Congo, but with lower cost.

Fig. 4.21 China-aided stadium in Ivory Coast: rendering(top); and the stadium in construction(bottom) (Source: top and bottom left, from BIAD; bottom right, from https://www.trendsmap.com/twitter/tweet/1152513481477906437).

National Wrestling Stadium of Senegal

China established diplomatic relations with Senegal in 1972 and China donated numbers of construction and other economic assistances to it ever since, including the national theatre, African civilization history museum and the children's hospital[①]. In 2014 the Senegal requested for a modernized wrestling stadium from China and was guaranteed by the donor with grant assistance in design and construction. The wrestling sport is regarded as one of the "national treasures" and beloved by the local people, who, as the Senegal government stated, had dreamed of one wrestling specialized stadium for ages. The stadium was used for holding wrestling competitions of African areas as well as other events such as gatherings and performances. Similar to other China's foreign-aided stadiums of the new era, the design job was appointed through a bid in domestic China. However, for this project and after, Chinese experts chose three alternative schemes while the representatives from

① http://www.xtrb.cn/news/2009-01/30/content_119835.htm.

the recipient country made the final decision of the winner. IPPR won the bid finally with their efforts to satisfy the recipient country in obvious architectural expressions of the wrestling sport.

Located in the capital Dakar, the 20,000-seat stadium covers 7.5 hectares areas with 18,000 m^2 gross floor area. Nevertheless, for Chinese architect, there was no precedent or reference for such a large professional wrestling arena before the construction of this 20,000-seat wrestling stadium. There are no regulations for constructing wrestling arenas in China's sports building standard, nor were there any construction drawings of wrestling stadiums in China's standard construction atlas that could be used as a reference. These conditions made applying Chinese or international standards to the design of this stadium difficult, and in some cases, adapting the design to these standards was impossible. Therefore, Chinese architects learned the local habits and used modern sports/stadium technologies in the way that considered local conventions. In fact, Senegal's traditional wrestling activities (called "lutte") originated from its harvest ceremony and the folkloric performance with a strong sense of ritual was one significant part. It had unique processes and rules that were different from those of international wrestling. The whole game process included admission, preparation through lining up, a folk show, warming up of the players and competition in the central area. According to the process, the oval field was determined to be 108 m × 68 m and centred by a sand-filled 20-m-diameter competitive venue where two wrestlers competed with each other. One preparation area, two warming-up areas and one performance area (which were also filled with sand) were located alongside this competitive venue. Since the ritual performance was also significant, ramps, made of hard plastic, for ritual folk custom shows and award presentations were arranged between the stands and the competition area to facilitate the interaction between contestants and the enthusiastic audience (Fig. 4.22). The size of the central field also considered beach football games. And a concrete floor beam was reserved below the 600 mm thickness of the ground surface to facilitate the temporary construction of a platform to realize the multi-functional use requirements of various performances and activities.

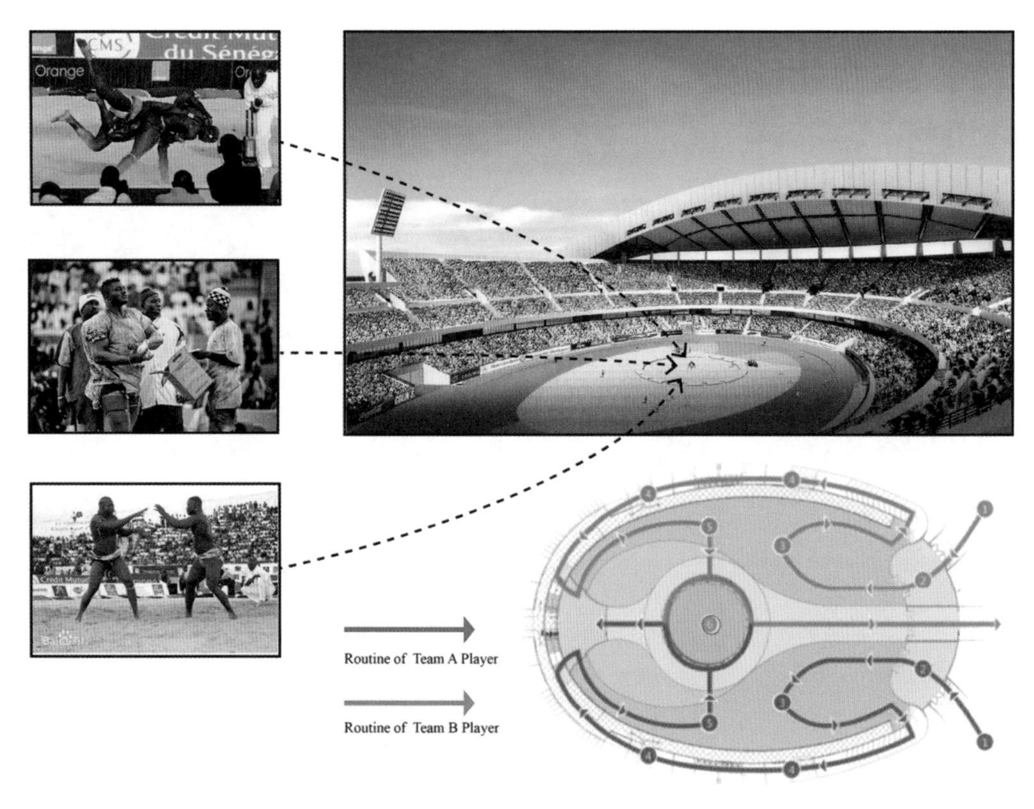

Fig. 4.22 Wrestling areas and the routines of the players of China-aided national wrestling stadium in Senegal (Source: drawn by the author based on the photos provided by IPPR)

The author believed that the reasons for the IPPR's victory in the bid lied not only in the satisfaction of the basic functions of the stadium, but more in the direct and slightly exaggerated expression of the local culture. The "golden belt" concept was used in the design of the roof structure to symbolize the wrestling sport and its unique culture of the country: a "belt" signature shape across one side of the stadium facing the main entrance to form the partial coverage of the U-shape stand. This roof structure was currently the largest steel architectural structure in Africa, with a horizontal span of 206.9 m, a maximum height of 50.46 m, and a width span of 47 m. In order to achieve the "belt" form, the axis of the rear arch was designed as a spatial curve, and a spatial triangular truss was used to resist the torsion caused by the design, since there are no additional supporting elements in the middle of the entire arch structure. Wind tunnel test and roof slab anti-wind test were carried out on the model of the stadium, according to which the purlin was consolidated, and the connection between roof slab and purlin was strengthened to ensure the safety of the roof (Fig. 4.23). The achieving of the "belt" form consumed the majority of the cost of the stadium,

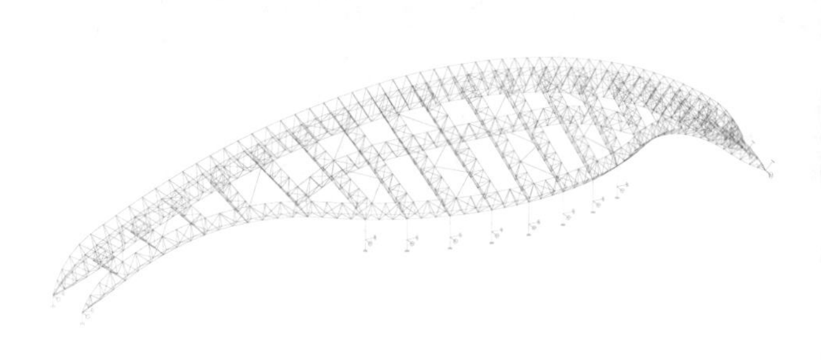

Fig. 4.23　The view from the main entrance of China-aided wrestling stadium in Senegal (top); and wind effect simulation of this stadium (bottom) (Source: provided by IPPR)

Fig. 4.24　Producing processes of the decorations in China-aided wrestling stadium in Senegal (Source: provided by IPPR)

with the simple frame structure for the remaining parts and the plain concrete as the material of the walls, which saved on the budget and improves easy future maintenance.

In the design of this stadium, even the decoration was designed to be more conspicuous in large scale to convey the respect of the regional characteristics from Chinese architects. To further express the wrestling theme, a 310 m-long, 4 m-high relief wall was hung above the entrance facade under the eaves of the "belt" roof structure. Five paintings of the scenes of their traditional wrestling games created by Senegal artists were later processed through 3D-animation handling by computer to form 54 pictures, which were made into relief models. Each picture was divided into 20 pieces based on which GRC (Glass-fibre Reinforced Concrete)-material-relief sculpture panels were produced in domestic China. The 1,080 pieces were transported from China to Senegal for on-site assembly (Fig. 4.24). Admittedly, Chinese architects overcame more difficulties to win the preference of the final decision maker (the Senegalese) of the bid and presented their concerns about the regional aspects in a more expensive, complex and high-profile way in architectural designs than that of the previous period, such as the Moi stadium.

Chinese enterprise completed the construction of the wrestling stadium in June 2018(FIg. 4.25). As the first modern wrestling arena in Africa, it also provided sports venues for surrounding schools and residents in addition to hosting regional wrestling games. The high-profile expressing of regional culture and characteristics in this wrestling stadium seemed to be the inherit of design approaches in the design of another China's foreign-aided stadium in Cambodia, from which IPPR was encouraged and gained experiences in winning the bid, as illustrated in the following case.

Fig. 4.25 Wrestling Stadium in Senegal after construction (Source: provided by IPPR)

New National Stadium of Cambodia (Morodok Techo National Stadium)

Cambodia, as one of China's best partners in the Asian area, has received a lot of China's assistance, such as the years of assistance in the maintenance and protection of the Chau Say Tevoda in Angkor Wat by Chinese experts since 1998 (Gu, 2004; Deng, 2018), and the loan assistance for road repairing after the 2011 natural disaster in Cambodia (Liu, 2019). China also became the country that sponsored the longest road repairing for Cambodia (Sun, 2012). China's BRI further strengthened its relationship with Cambodia, making Cambodia one of the Asian countries that received the most construction aid from China.

In 2014, China agreed to finance one large stadium with a high cost of RMB 0.925 billion yuan. It was China's most costly, largest-scale and highest-standard foreign-aided stadium, as well as being China's most expensive foreign-aided construction projects. Different from China-aided stadiums of the previous period that emphasized on economic efficiency, this stadium might equal the large stadiums of China's major cities and would be qualified for holding continental sports events, big international football games and national activities, as the new national stadium of Cambodia.

In fact, like many other China's foreign-aided stadiums, this new national stadium was the main stadium of Morodok Techo National Sports Complex in Phnom Penh, a 94-hectare sports centre approved by Cambodian Prime Minister Hun Sen in 2012, as the country's first "modern multipurpose and international standard sports facilities" (Ponlok & Chamroeun, 2012). This sports complex received much attention and budget support (around 200 million dollars) from the Cambodian government, being the main venue of the 2023 Southeast Asian Games (SEA Games) (Vorajee, 2019).

This sports complex was located at Prek Tasek Commune in the capital's Russei Keo district, between the Tonle Sap River and the Mekong River (Fig. 4.26). The construction was divided into three phases. Phase Ⅰ lasted from 2013 to 2017, including the construction of the indoor sports centre, training halls, aquatic centre, athletes' village, and training fields, which cost USD 38 million and were solely financed by the Cambodian government and built by the local firm L.Y.P. Group, which owned the land. Phase Ⅱ mainly included the main stadium, hockey fields, traditional sports hall, and gymnastic centre from 2017 to 2019, which cost USD157 million and the main stadium was financed by the Chinese government. Phase Ⅲ consisted of health care facilities, sports medical centre and tennis centre and was expected to be completed in mid-2021.

Cambodian architect firm Architect Solution Company was mainly responsible for the

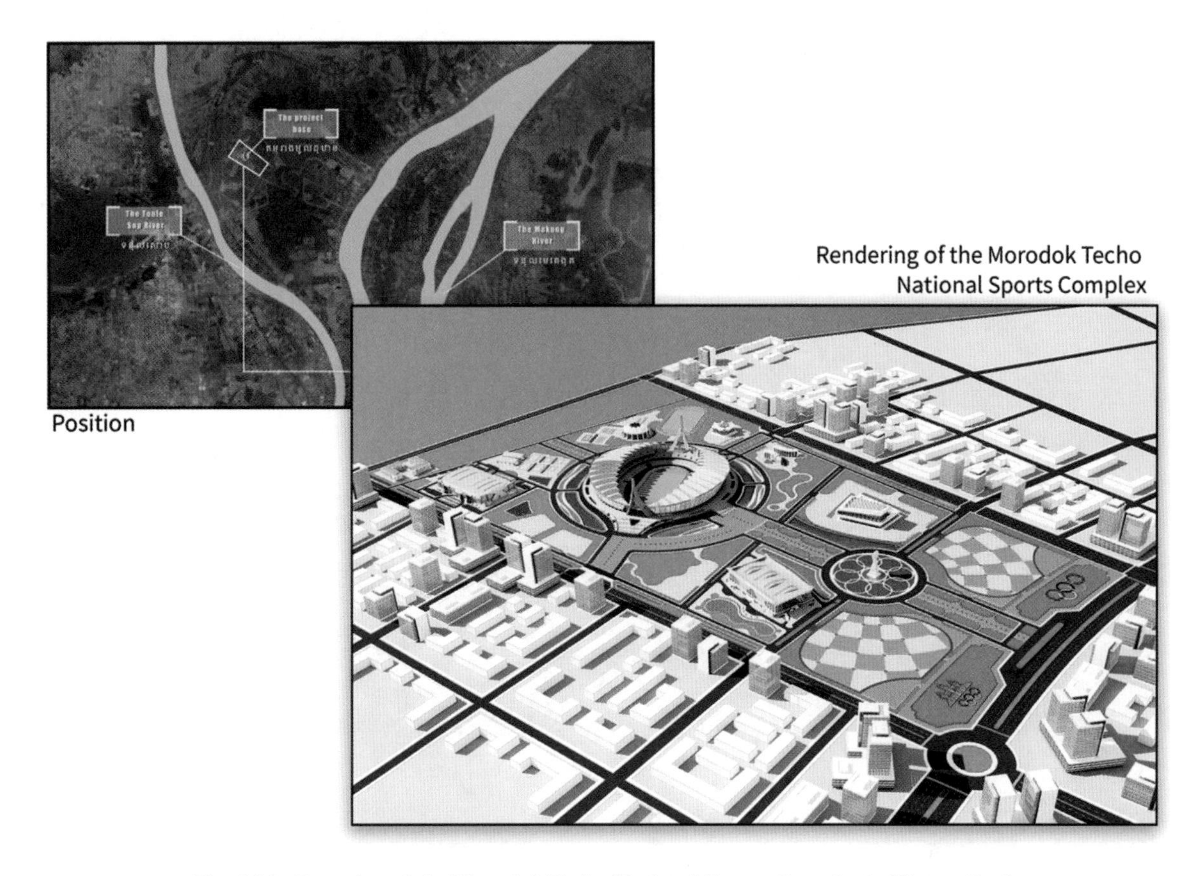

Fig. 4.26 Location of the Morodok Techo National Sports Complex in Phnom Penh
(Source: top, provided by IPPR; bottom, http://www.mtn-stadium.com.kh/?detail=37&ip=37&title=Master-Plan-ប្លង់សរុប-Master-Plan)

initial design and layout planning of the complex as well as the stadiums other than the main stadium. As explained by the principal architect of Architect Solutions Company[①], Mr. Varheng Dawuth, the design of the sports complex was intended to utilize locally-sourced building materials and building workers as much as possible for economic reasons. The overall plan of this complex was inspired by the architectural history of Cambodia. The water and road systems of the layout were also inspired by the structure of the Angkor Wat Temple. The individual building's design was inspired by Cambodian cultural features such as the Angkor Wat and Banteay Srey Temple and the symbols of dragons and elephants (Fig. 4.27).

① https://www.construction-property.com/born-from-khmer-art-to-meet-global-standards/.

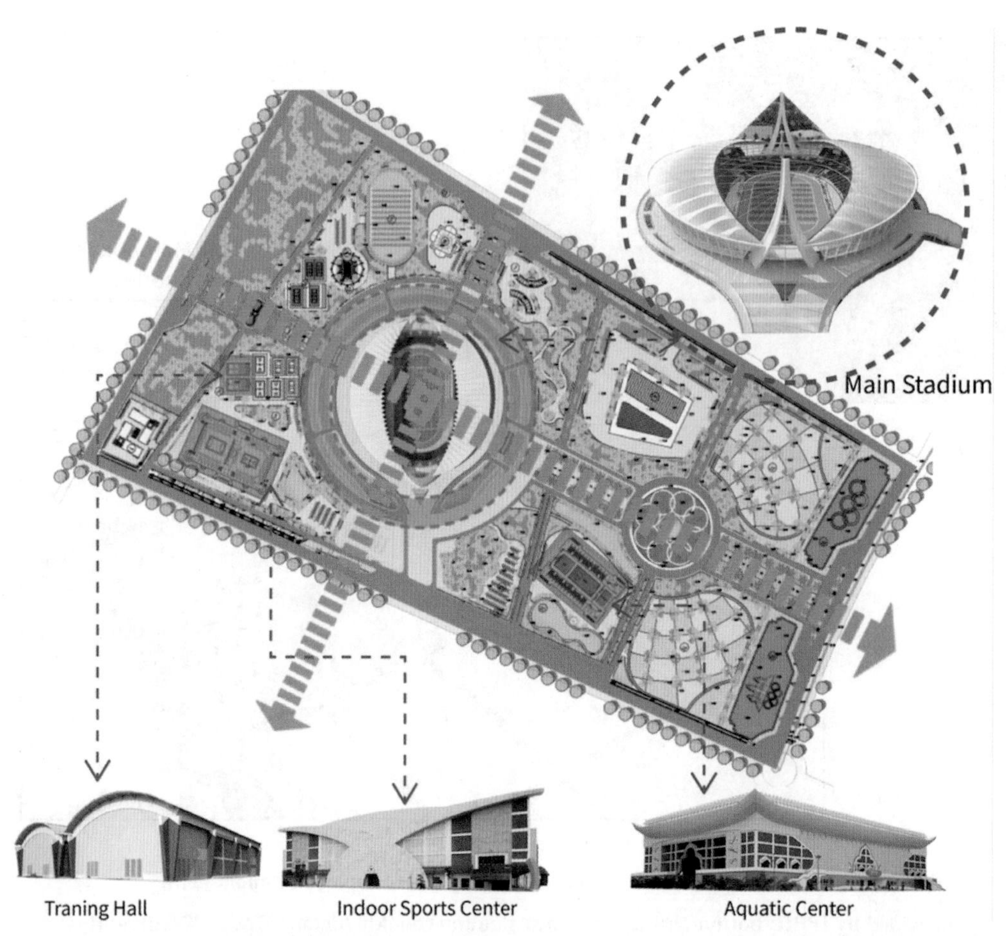

Fig. 4.27 Layout analysis with individual stadiums of Morodok Techo National Sports Complex in Phnom Penh. (Source: drawn by the author based on the pictures from http://www.mtn-stadium.com.kh)

For the main stadium aided by China, the design was completed by Chinese design institutes through a bid in domestic China. The bid evaluation by Chinese experts selected three design schemes in 2015. Later in April, a special meeting about the final selection of the design scheme of this stadium was held in China's department of Foreign Economic Cooperation[①]. In response to the Cambodian willingness, Chinese government broke the routine and discussed with the recipient country before the approval of the projects. The alternative schemes were introduced to the representatives from Cambodia, which reflected their concepts, materials and maintenance, etc. However, the representatives could not decide, so they reported the schemes back to the governmental authorities of their country. By

① http://www.mofcom.gov.cn/article/shangwubangzhu/201504/20150400941952.shtml.

the end, it was Prime Minister Hun Sen of Cambodia that made the final choice upon his preference on the design from IPPR①. This stadium became the first China's foreign-aided stadium that the design was chosen by the head of the state of the recipient country.

According to the general layout of the sports complex, the stadium was set to be located at the centre, separated from other areas by a circular wading pool, and connected with them in series via the axes in four directions to form a "one ring and two axes" layout (Fig. 4.28). Covering 14.9 hectares' site areas, the main stadium accommodates 60,000 seats with 80,000 m^2 gross floor areas. Surrounded by the indoors stadiums designed by Cambodia's local architects that shared unique features symbolized the country's tradition and culture, the imported mega-structure needed to be accommodated with the existing local contexts. However, as the largest and most modernized stadium of the country and being designed by the donor's state-owned design institute, the architectural expression seemed to be obviously different.

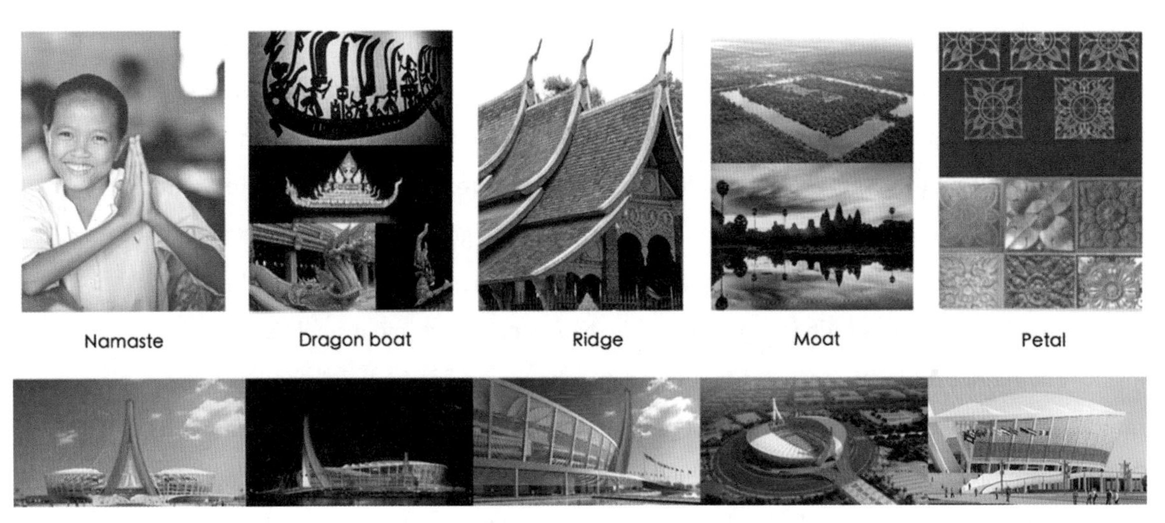

Fig. 4.28 The design concept of China-aided Cambodia's new national stadium (Source: redrawn by the author based on pictures provided by IPPR)

As explained by the architects from IPPR in the interview of the research, the design scheme from IPPR won Hun Sen's② preference for its adoption of multiple regional design approaches of cultural expression combined with modern architectural language and tech-

① China IPPR International Engineering Co., Ltd won the bid of designing and managing of this China-aid stadium in Cambodia.

② Hun Sen, the premier of Cambodia, made the final decision among three design schemes provided by Chinese government after the primary choice by Chinese experts in the bid of designing the new national stadium.

nologies. The form of the stadium followed the roof ridge of traditional Cambodian buildings, as two giant bridge columns with unique shapes were set at both ends of the north–south axis of the main stadium, serving as the convergence support of steel cables of its roof membrane structure. The front appearance of the giant columns simulated the hand gesture of "namaste", which was a traditional Cambodian and Buddhist greeting, while their inclination coincided with the side shape of the structure to symbolize a dragon boat, a significant culture element of Cambodia. Such imitation also existed in macro and micro scales, as such the ring-shape water system was located around the main stadium in the layout planning, as a reflection of Cambodia's traditional planning idea of "moat" (Fig. 4.28). Furthermore, the decoration of special flower-shaped patterns was attached in the hollow grid plates of the facade coloured gold, as Cambodia's favourite traditional colour.

The use of the concepts of "namaste" and "dragon boat" was favoured by Cambodia's leader but generated challenges for architectural techniques, especially the structure and construction. It was achieved through a complex structural design with unique hyperbolic herringbone-shape towers, a large-angle ring column cable membrane truss awning system and a ring column-beam supporting structure system. The section of the herringbone-shape tower was gradually reduced from the bottom to top and closed at the height of 78 m, finally reaching a height of 99 m. The top height of the tower was formally designed to be 96 m but later changed by Hun Sen to be 99 m for his belief of the good fortune behind the new number. The exterior of the tower was made of fair-faced concrete, while the middle body was hollow with partitions set horizontally, wing-shape steel bones and a steel bar skeleton inside. The 65-m cantilever cable-stayed locking membrane awning took the ring beam and ring columns as the support system, and the ring columns tilted outward at an angle of 67 to 79 degrees from the ground. BIM technology was used for the steel mould configuration of the beam and columns, and SAP simulation analysis was carried out to control the tension deformation. ETFE was used as the membrane material, which had been widely experimented successfully in domestic China. The metal curtain wall composed of stainless-steel cable and 1.2-mm-thick aluminium-magnesium-manganese perforated plates covered between the ring columns around the stadium as the facade. ANSYS software was used to analyse the cable tension and determine the tension sequence (Fig. 4.29). The landmark and symbolized form was achieved with the help of advanced computer software and materials by Chinese technicians.

Additionally, climate-concerned regional designs were utilized in the stadium, as many

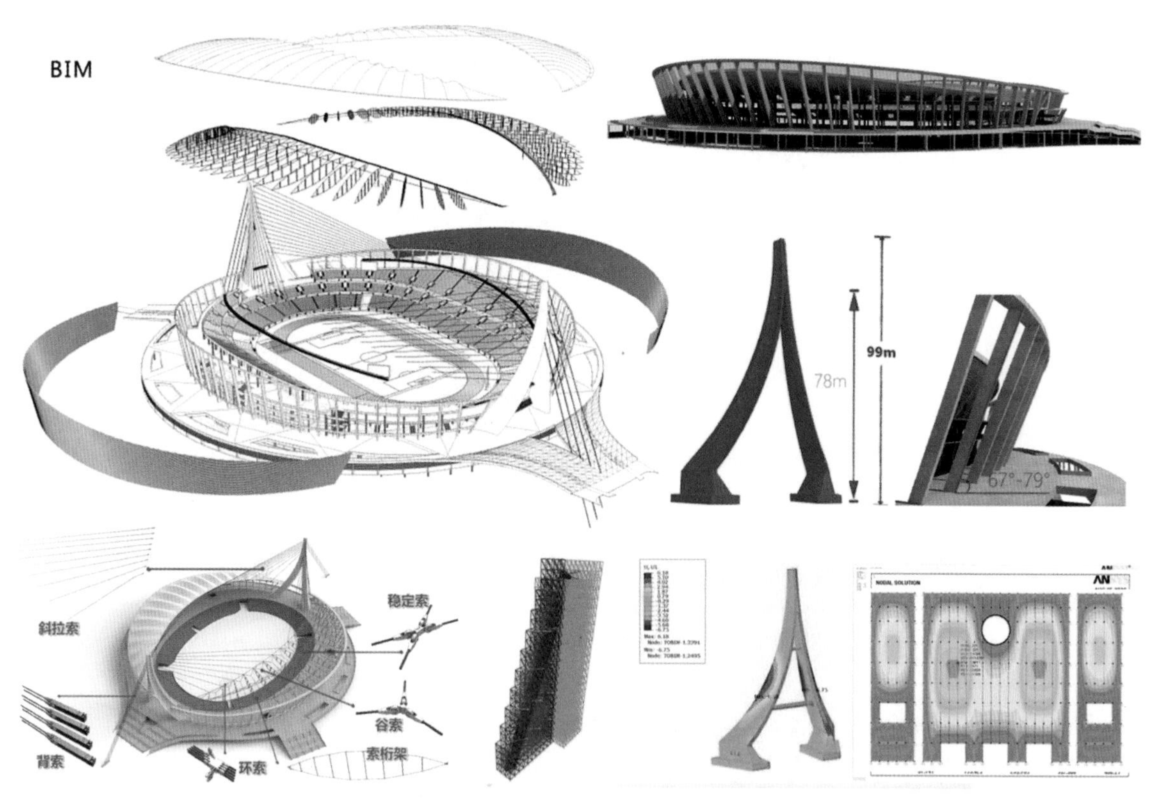

Fig. 4.29 Structural analysis of China-aided Cambodia's new national stadium (Source: provided by IPPR)

previous China-aided stadiums did. To cope with the hot dray season and long rain season in Cambodia, plenty of open space was reserved under the first-floor platform, forming an all-weather activity venue. The main facades of the stadium adopted aluminium perforated plates, and the awning used a membrane to meet the requirements of sheltering from sun and rain and natural lighting and ventilation. The hollow plates also contributed to the natural ventilation of the stadium. All the three-layer stands were overlapped to introduce more air flows. The under-back areas of some seats were hollowed for ventilation and cooling, similar to the designs in the Olympic Stadium of Cambodia in the 1960s (see Chapter 2) (Fig. 4.30). The surrounding water system improved the microclimate of the site, which was also similar to the design of the Olympic Stadium of Cambodia. Chinese architects had gained experienced in using these passive-efficient technologies through years of being involved in the foreign-aided projects, and through the knowledge learned from China-aided stadiums of the early period.

Fig. 4.30 Climate-concerned design approaches in China-aided Cambodia's new national stadium: (a) natural ventilation through under-stand space and the hollow aluminium perforated plates; (b) the hollowed under-back areas of some seats; (c) the general natural ventilation around the stand areas (Source: provided by IPPR)

Due to the sufficient budget, and high-standard requirement from the recipient country, the design of the stadium had a greater sense of internationalization, luxury and commercialization. The field layout met the requirements of field, track and football competitions and standardized sports techniques used were qualified for international standards such as the IAAF and FIFA. An international standard emergency exit system was applied in the layout design of the stadium. Arranged in a saddle form, the three-layer bleachers accommodate 55,000 seats with a platform reserved at the north and south for 5,000 temporary seats. A wide evacuation platform was built to allow the audience to evacuate in 7 minutes and 7 seconds. Walkways for the spectators were located in the north and south sides, with large ramps leading directly from the ground floor to the first floor. The east and west sides were mainly for special pedestrians and vehicles, to the entrance on the ground. Compared with previous China's foreign-aided stadiums with the economic layout of the under-stand space, this stadium aimed to integrate more diversified functions such as commercial and entertaining ones. Even the interior design showed more improvement with more fancy and completed designs through BIM+VR techniques (Fig.4.31).

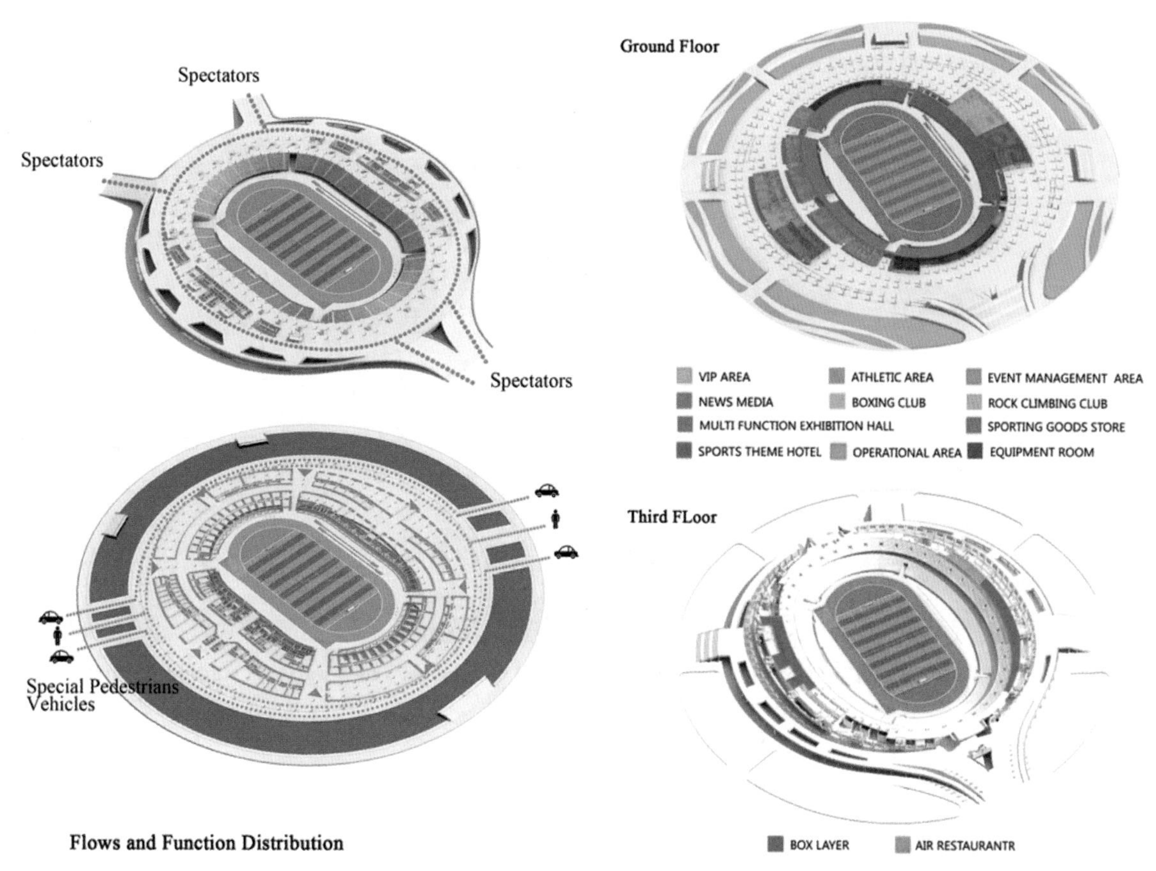

Fig. 4.31 Function layout of China-aided Cambodia's new national stadium (Source: provided by IPPR)

The construction began in June of 2017 and was completed in July 2021. The Chinese firm was entirely responsible for construction, partnered with L.Y.P Group as the developer, with 80% local labours and 20% Chinese labors. The principle technicians were mostly from China, same with the steel materials and equipment. The concrete and sand materials were bought locally. A concrete batching plant and a prefabrication factory were set up on-site for field production of the units for the stands. This was the first time that the BIM was utilized in a China's foreign-aided construction projects, and it had been increasingly wide-spread in the design and construction of Chinese domestic projects. Chinese enterprises started to experiment new techniques in overseas projects for profits and efficiency with more open attitudes authorized by the government of the new period. (Fig. 4.32, Fig. 4.33)

Fig. 4.32 The site planning and bird view of China-aided Cambodia's new national stadium (Source: provided by IPPR)

The accepting attitude by the recipient country reflected the country's expectation on stadium projects to be symbolic and regional. For winning the bid, Chinese architects tried hard to integrate more elements to reflect regional characteristics in aspects of the layouts, facades, forms, details, decorations and techniques. Sometimes the prospect of modernization of the recipient country diverged from the understanding of Chinese architects, and its appealing influenced the design to be out-timed to some extent. Interestingly, the preference of the "namaste" from Cambodia also appeared in the choice of another China's foreign-aided project, one public hospital in Phnom Penh which had the similar obvious form symbolization of the welcome gesture, also designed by IPPR (Fig. 4.34).

Fig. 4.33 Construction Processes of China-aided Cambodia's new national stadium (Source: top two and bottom left, from IPPR; bottom right, the photo was taken by Guo Xiaofeng in 2019, in which the author was on a site investigation with Chinese engineers Mr. Liu Baiqing and Mr. Yin Zhiwei from IPPR)

Fig. 4.34 China-aided hospital in Phnom Penh (Source: provided by IPPR)

4.4 Summary: Bi-influences on Chinese Architects' Tables

The new millennium witnessed significant transitions in China's foreign-aided construction projects, as well as the stadium projects. The number of the stadiums experienced explosive growth and the geographic distribution varies in fourth continents (Africa, Asia, Latin Africa and Oceania) of the first decades, but returned to focus the Africa and Asia after the BRI proposed. In addition, the design qualities were also improved due to diversified financial supporting mode and a higher-level of requirement from the recipient countries.

For this new period, the reforming of China's construction aid policy and the mechanism was significant, which generated considerable and obvious influence. The introduction of bid system in the foreign aid project attracted more Chinese design institutes and architects to the designs of China's foreign-aided stadium projects, especially for the first two decades of the 21st century. The allowance of cooperation with foreign/local enterprises by Chinese government enriched the designs of these stadiums, and the more diversified financial supporting modes provided relatively sufficient budget for higher-standard stadiums. And the invitation of representatives from the recipient countries as the decision-makers forced Chinese architects to care more about regional aspects in their creations. Therefore, the influence generated by policy and mechanism of this period seemed to be the most apparent and greatest of all three periods.

Economic aspects seem to always be constraints for foreign aid projects, and the new century is no exception. A new mode of financial supporting enables more budget for the stadiums, which makes the economic limitation to be less influential. Known for the interview of the research, some projects experienced the funding shortfalls and got extra supports through the negotiation between the two governments, which seldom happened in previous stages. However, it should not be ignored that these projects are still aided projects exported into less developing regions rather than purely commercial ones, not to mention the advanced stadiums of the developed regions. The attribute of being gifts requires Chinese architects never forget to consider the economy into their minds when completing these special missions.

For the function aspects, these overseas stadiums hold relatively simple and basic functions in the initial years of the new century. However, later in the recent decade, with the development international stadiums (as well as Chinese domestic stadiums) which has been more commercialized and multi-functional, the recipient countries require their stadium gifts to be up-to-date, especially when they see stadiums or sports complexes as the accelerants of the regional development. Performance, business, commercial service and other functions are filled into these overseas stadiums, and the post-game operations need to be considered. Such a trend is basically synchronized with the development of China's domestic stadiums. Therefore, the influence of functions on the architectural design of China's foreign-aided stadiums has been raised to a higher degree.

China's preparation for the Beijing Olympic Games boosts the development of Chinese domestic stadiums with the importation of foreign architects' designs and construction of stadiums, the construction of high-standard stadiums, and the blooming of sports facility development of China's other cities under the post-Olympic influence. Such development generats impacts on China's foreign-aided stadiums, such as the "skin" popularity, the layout and safety designs. However, it should be noted that the Chinese design institutes that have long been active in China's foreign-aided stadiums are not actively involved in the designs of Chinese domestic stadiums. Besides, their designs of these overseas stadiums develop and transform based on their previous designing routines according to the rules of simple formal beauty. The focus on regionalism, the simple forms and pure style make these stadiums distinctly different from China's domestic stadiums.

As mentioned in the previous paragraphs, the distribution of these stadiums spreads to more various regions in the 21st century and receives more attentions and inputs from the recipient countries. As part of this, more concerns about the climate are requested in the designs of these stadiums, such as the use of materials, the layout and the section designs. Especially for the stadiums in hot areas such as in Africa and Southeast Asia, climate aspects play significant roles in architectural design. In some high-standard stadiums, the climate-adaptive approaches used not only low passive technologies but also active high technologies. Compared with other factors in this stage, the influence of climate is relatively higher; but if compared with cultural aspects, it may be slightly lower.

In the 21st century, the symbolization of the local culture has been welcomed by the recipient countries, for the increasing national awareness of the recipient countries. The increasing national and cultural awareness of the recipient countries leads to the requirement of more expression and symbolization of their culture and nation in designs, especially when the recipient countries sought hold more discourse rights in the decision process for the design scheme. This requires Chinese architects to concern more about appropriated regional design approaches in architectural design to satisfy the new party, of which the cultural expressions and metaphors tend to be the protagonists planted into the architectural design in a new way under the discourse of international style, with the combination of new structures, materials and technologies. In the new period, cultural aspects' influence on the architectural design is fundamental and forceful and has become the main origins of the conceptual designs of numbers of China's foreign-aided stadiums.

One of the most obvious changes of this period is the increasingly more involvement of the recipient country in the design of these stadium projects. The role of the recipient country transformed from no interference in the previous two periods to certain opinions on the scheme design or modification, sometimes even completely overturning the previous design by Chinese architects, and becoming the final decision-maker in the bidding in recent years.

In this section, architectural technique aspects are analyzed in the end because of its relationship with the previous factors. The higher standard, larger-scale/span cultural expressions through symbolization in these stadiums require more advanced architectural technique support, especially in recent years. In fact, since the 1990s, China's domestic stadiums have experienced significant development on the structure, material and construction, and was once again promoted by Beijing Olympic Games. Such development enables Chinese technicians to satisfy the architectural technic requirement in China's foreign-aided stadiums with complex structure designs, high-tech materials, and advanced simulations in the computer software. It is believed that the influence of this aspect is higher than that of previous periods but remains at a medium level.

Generally speaking, in the 21st century, significant transitions have happened in the influential aspects of the design of China's foreign-aided stadiums. China's foreign aid policy and mechanism generate more impacts on the designs, mainly for the utilization of tender and bid system, and more diversified fund modes in these overseas aid projects, with the economic aspect thereby decreasing in impacts. Another significant change was the increasing participation of the recipient countries in deciding the design due to the transforming of the mechanism, leading to the "recipient countries' opinions" to be one of the most influential aspects. Chinese architects considered more about the regional aspects and the recipient countries' preference in their designs. Cultural symbolizations become the main theme of the design concept and play critical roles in the designs of China's foreign-aided stadiums, especially in recent years. And the climate-orient approaches have been improved from low passive techniques of the previous period to high-tech ones of the new era. Regional aspects turn to be the main course on the design table of Chinese architects.

China's foreign-aided stadiums bloomed with diversities and transitions in recent years, improved in the design level of such stadiums with the pursuing of high-standard stadiums by the recipient countries and the great development of China's stadiums from pre-Olympic period to post-Olympic period. The gap between China's foreign-aided stadiums and its domestic stadiums is narrowing down in scales, functions, techniques and modernization levels. Although these stadiums cannot be considered as the best Chinese stadiums, they are mostly the best stadiums in the recipient countries. Influenced by China's special construction aid mechanism and the recipient countries' opinions, these stadiums have become the productions of bi-choice of the donor and the recipient sides. The author considers these designs of the new era to represent a unique critical regionalism with cultural and climate elements embedded in the mega-structures with modernism, adaptation, technique, culture or combination thereof. Such regionalism differs from China's domestic stadiums and speaks for itself due to its particular attributes. Still, it cannot be established without the basement of the development of Chinese domestic stadiums. The relations between them will be further discussed in the final chapter of this book.

CHAPTER 5
CONCLUSIONS

5.1 From Regionalism to Critical Regionalism: the Development of Design Approaches

Since the initial period, China's foreign-aid construction projects have given Chinese architects valuable opportunities to experiment with their ideas and express their understanding of modernity. However, due to the design level of China's domestic stadium, most of China's foreign-aided stadiums of the first period tended to be copies of China's existing domestic stadiums, with only a few projects attempting to consider the locality.

Although China's overseas aid stadiums of the second period tended to be the conventional design with low standards and limited cost, due to the adjustment of China's aid policy, improvements were made for the regional designs. The cost-effective low-passive techniques had been familiar to Chinese architects and were commonly utilized in the designs of China's foreign-aided stadiums. Also, more cultural expressions in the detail design can be found in some representative cases of the stadiums in this period. The author regarded them as the appearance of more basic regional concerns in the designs by Chinese architects.

After 2000, the recipient countries that succeed their independence from long-time colonization in the 20th century or even in the 21st century, prefered to express a strong consciousness of nationality and culture in their stadiums. The reform of China's foreign aid mechanism introduced more inputs from the recipient countries, which forced Chinese architects to integrate more regional approaches in their works to satisfy the local. With the development of architectural techniques, the internationalized standard requirement, and the sufficient supporting budget, these gift stadiums were supposed to be modernized but also feature local characteristics. Regional designs were planted into the architectural design of the stadiums in a new way under the discourse of international style, with a combination of new structures, materials and technologies. And the author proposed that such approaches stepped towards high profile ones and thus transmitted from the simple regionalism to critical regionalism.

Generally, this study concluded the regional design approaches in China's foreign-aided stadiums into three categories, climate-adaptive approach, culture-oriented approach and standard-adaptive approach. The period, background, effect and representative stadium cases of the three categories have been explained briefly and respectively in Table 5.1.

Table 5.1 Regional design approaches in China's foreign-aided stadiums

Regional Design Approach	Climate-adaptive Approach		Culture-oriented Approach		Standard-adaptive Approach
	Low-passive	High-tech	Decoration	Symbolization	
Period	1950s–1970s; 1980s–1990s	2000s-	1980s–1990s; 2000s-	2000s-	2000s-
Background	Early aid mode; economic limitation	New financial supporting mode	Economic limitation; discourse right from the recipient countries	Tender and bid in aid projects; discourse right from the recipient countries	Entering the market
Effect	Save the budget; be adaptive to the local climate; energy saving;	Be adaptive to the local climate; energy saving; express modernity	Satisfy the recipi-ent countries	Help win the bid; Satisfy the recipient countries	Neutralise differences between standards, customs and habits
Representative Case	Stadiums of Moi International Sports Centre, Kenya	Tanzania nation-al stadium; Cambodia's new national stadium	China-aided Stadiums of Moi International Sports Complex, Kenya	The China-aided stadium in Senegal; Cambodia's new national stadium	The China-aided stadium in Senegal; The China-aided stadium in Costa Rica

The climate-adaptive approach has been used from the early period to the present, with the low-passive ones in the first two periods and the high-tech ones in the new era. Low-passive techniques were encouraged in the designs of stadiums in hot climate areas for natural ventilation and sunshine shadowing cost-effectively. While high-tech approaches only appeared in recent decades using advanced technologies such as new advanced materials and mechanical systems to achieve better functioning of the stadium regarding the local climate with modernity expressions.

The culture-oriented approach of China's foreign-aided stadiums originated from the decoration designs such as the local statues or other detailed elements of the facades or interiors, to gain some expressions of the local culture. These approaches were generated due to the design level, economic and architectural technique limitations. The outcomes were normally welcomed by the recipient country so that they are continually used in China's contemporary overseas aid stadiums, but with more advance materials or techniques.

In the new era, cultural expressions in the designs of China's foreign-aided stadiums were more symbolized, for the expectations from the recipient countries on the stadiums to be the symbol of their respective nations, as normally the gift stadiums from China were the first modern sports venues (or the first high standard and modernized stadiums) built in these countries and designed for holding international sports events. As Cassirer (1983) mentioned, "no··· longer in a merely physical universe, man lives in a symbolic universe." The symbol can be regarded as a clue to the nature of man. Thompson even used "symbolic conception" to describe culture[①] (Thompson, 1990). In architecture, symbolization has been commonly utilized in form designs from ancient times to the postmodernism period as a clear approach to express local characteristics and to add regional attributes to the buildings. By displaying the characteristics of certain national objects, such as form and colour with abstract metaphor or imitation of the culture related articles, China's foreign-aided stadiums (especially the lately stadium projects) can become a colossal new totem of the national culture of the recipient country.

Here, the standard-adaptive measures in the designs of China's foreign-aided stadiums, which appear in the 21st century, are also listed as another significant regional design approach in the research. As mentioned in previous paragraphs (Chapters 2, 3 and 4), it is the compulsory requirement by MOFCOM to use Chinese standards in the designs (as well as construction). Chinese architects need to adapt their familiar standards into the local circumstances—otherwise, conflicts and problems may happen. Some participants pointed out in the interviews that at present, they are supposed to learn all the related buildings codes of the recipient country before their design work for the preparation of persuading them to accept the Chinese design. This type of regional design approach is believed to only exist in overseas constructions, especially the aid constructions when the donor supports the finance and uses its own standards and products in return. Chinese government is experimenting with the new "localization" mode in some of its foreign-aided constructions which encourage using the standard of the recipient countries or offer them the opportunities of designing and constructing the projects under certain management from Chinese enterprises. Although such reforming has not spread to the stadium projects, the author believes that the standard-adaptive approach will be greatly influenced by it in the near future, as will be discussed in the following section.

Basically, from the analysis results of the research, it can be concluded that the regional design approaches in China's foreign-aided stadiums appeared in the initial period, gently

① Culture is the pattern of meanings embodied in symbolic forms, including actions, utterances and meaningful objects of various kinds, by virtue of which individuals communicate with one another and share their experience, conception and belief.

developed in the second period and bloomed in the new period. After 2000, the architectural designs of China's foreign-aided stadiums have a better design quality, which considers the regional, cultural and local contexts, with the international style and cultural-symbol elements combined, with the modernist and regional design approaches combined (Chang & Xue, 2019a).

5.2 Similarities and Dissimilarities with Chinese Domestic Stadiums

The period of this study is divided according to the significant changes in the development of China's foreign aid. Such division is actually different from the development stage of China's domestic stadiums, which received significant development acceleration by the two large sports events held by China (Beijing Asian Olympic Games in 1990 and Beijing Olympic Games in 2008), reflecting the inseparable relationship between the sport and stadiums. China's foreign-aided stadiums, by contrast, tend to be more affected by aid policies and mechanisms. By comparing similarities and dissimilarities, the complexity of Chinese architectural history can be reflected (Fig. 5.1).

Similarities exist in the development paths of China's domestic and foreign-aided stadiums, with parallels and intersections, but dissimilarities occupy the majority. It is very interesting to see the importation of the techniques and experiences gained in China's foreign-aided stadiums to China's domestic stadiums of the early period, and later the reversed exportation from China's domestic stadiums to China's foreign-aided stadiums. At the same time, it must be conceded that the star architects and some Chinese design institutes who have been actively involved in the creation of large-scale noted stadiums in domestic China are not qualified by the Chinese government to participate the design of China's foreign-aided stadiums. The design level and features of China's foreign-aided stadiums to be relatively conservative. All the above have contributed to the separations of the development of China's foreign-aided stadiums from the domestic ones since the second period.

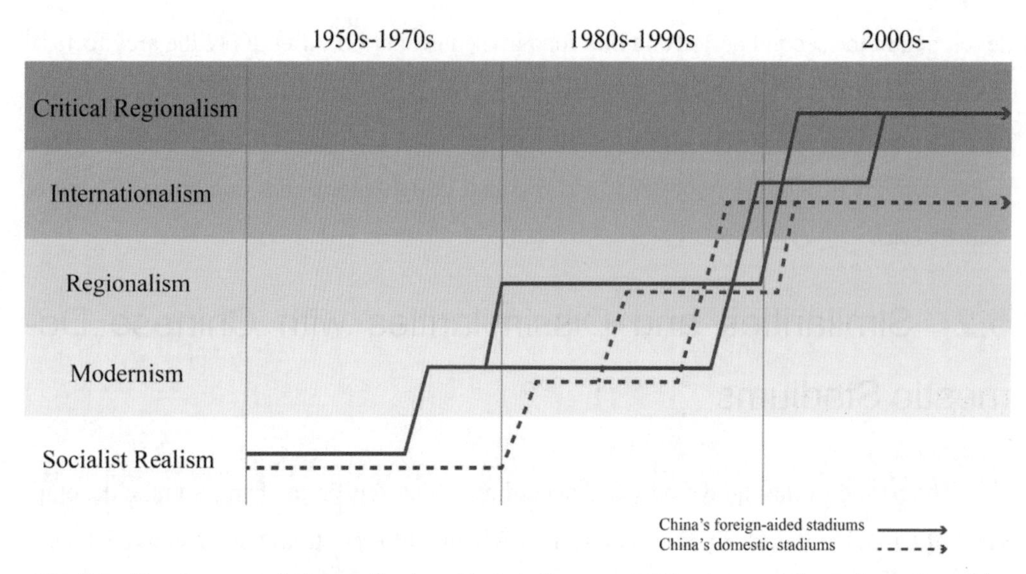

Fig. 5.1 Development of China's foreign-aided stadiums and China's domestic stadiums in architectural designs (Source: drawn by the author)

Meanwhile, due to the different cultures, regions and development degrees, the pursuit and understanding of modernism in each recipient country are different from each other and also different from that in Chinese architecture. Regionalism has gradually joined in the designs of China's foreign-aided stadiums since the 1980s, and the reform in China's foreign aid mechanism introduced greater participation of recipient countries, which has strengthened the proportion of regional design, and also has led to the obvious differences between China's foreign-aided stadiums and its domestic ones. The exportation of modernism with critical regionalism and exotic features constitutes a special stroke in the development history of Chinese sports architecture.

Since the 21st century, with the rapid development of China's domestic stadiums generated by Beijing Olympic Games, numerous high-standard and widely discussed stadiums have been designed and built in domestic, with new concepts, new architectural languages and new technologies. Some of the new development has been exported to China's foreign-aided stadiums, which helped China's foreign-aided stadiums catch up with China's domestic ones. However, they developed in different ways,with the former stepping into critical regionalism while the latter into internationalism.

5.3 Forecast of the Future Designs

The architect responsibility system is likely to be experimented in both China's domestic constructions and its foreign-aided ones, with the architect leading the whole process. The joint bidding body is emerging in China's foreign-aided construction, which conducts the design and construction management as a consortium bidding. And large design groups (e.g., some design institutes in Shanghai were merged into the ARCPLUS) also emerge with the consolidation of design institutions of the near regions. All these changes will generate profound influence on China's foreign-aided stadiums, as well as other foreign-aided construction projects. As mentioned in the previous sections, the "localization" mode is being tried, which may once again bring great reforms and even fundamental transformations to China's foreign-aided constructions. Instead of China taking full responsibility for all aspects of the design and construction, in the new mode, China is only responsible for financial support and some technical supervision, and the design and construction works are left to the local authorities of the recipient country. If the application of the "localization" mode gradually expands to all China's foreign-aid projects, Chinese design will disappear from these aid stadiums, which makes this research constitute a relatively complete part of China's foreign-aided stadiums designed by Chinese architects.

In fact, Chinese design institutes and Chinese architecture have tried to start the design of some overseas commercial construction projects in recent years. Some less developed areas still need the paid technical services from Chinese enterprises. Such commercial help is not included in this study as its commercial attribute is fundamentally different from the designs of China's foreign-aided construction projects. However, the aid activities provide experiences and knowledge in some aspects for the business projects abroad and prepare Chinese design institutes for their entering the international market. For stadiums project specifically, Chinese design institutes participate in only a few commercial stadium projects overseas (for example), and mostly, Chinese enterprises participate in the construction and management process of the commercial projects. In the 21st century, more commercial projects will enable Chinese design institutes and engineering groups to export their services and productions and change the situation of solely relying on the aid projects. The author believes that the development, transformations and differences among China's foreign-aided projects and China's foreign commercial projects deserve further attentions in future.

5.4 Concluding Remarks

In the very beginning, China's foreign-aided stadiums were designed by local architects with a reputation in the receipt countries, when Chinese architects and technicians mainly served as assistants, with China's exported financial, construction and labour supports (e.g., early China-aided stadiums in Indonesia and Cambodia). After the early 1970s, Chinese architects started to be involved directly into the design of aid sports buildings. These exported stadiums involved serious consideration of China's culture and philosophy of "sharing and giving". Such donating of best stadiums abroad provided good opportunities for Chinese architects to learn and improve their design abilities, which allow these China-aided stadiums to take the lead over the domestic ones. Although only the state-owned core design institutions participated in the designs of China's foreign-aided stadiums, Chinese architects seemed to assume the aid projects as great chances to design more freely and experimentally. Although the structure and technology of stadiums during that age were relatively simple, normally the overseas projects showed better design and construction qualities.

After China's reform and opening up, the change of diplomatic policy and the prominent increasing number of China-aided stadiums required more economical designs. Design institutes attending this field gradually changed from designated national core institutions to local ones from various provinces of China. Architects started to use the conventional methods to design cost-effective "gifts", in which low passive technologies were commonly used such as natural ventilation since most of the regions that received China's stadium aid were in the tropical climate. Furthermore, some improvements were achieved with certain basic regionals concerns to better adapt to the cross-border contexts.

In the 21st century, contemporary stadiums have tended to use new structures, new materials and new technologies, especially those closely related to sustainable and green technologies. Also, China's domestic sports buildings bloomed with massive large-scale and high-tech stadiums designed and constructed in significant cities, which in return promoted the design and construction of China-aided stadiums to follow the trend. Another significance of the new era is the multi-cooperation or competitive relations with other firms or even foreign firms due to the bidding operation of China-aided stadium projects. Such cooperation and competition contributed to the internationalization and modernization of these stadiums to some degree. Profound improvement of China's aid sports buildings was achieved to meet a high standard of modern sports buildings in the new era to hold international games, which may not have happened without better financial supports from diversiform financing models rather than simply donors and low-interest loans. Besides, the new emerging bid operation for aid projects have forced Chinese designers to plant more cultural concepts into the design for persuading the recipient countries, who stand as one of the main decision-makers. With China's promoting of the BRI, more overseas stadium projects will be planned and settled, which may hold a larger capacity and higher standard. In addition, China's foreign-aided stadiums consequently bloomed with more diversities and transitions of regionalism, modernism, adaptation, techniques, cultures or combinations of the above.

Here a summary is made of the architectural development of China's foreign-aided stadiums of the three periods, from aspects of policy, designer/institutes, working mode, function, style, structure, technology, etc., to form comparisons (Table 5.2).

Table 5.2 Summary of China's foreign-aided stadiums of the three periods

Period	1950s–1970s	1980s–1990s	2000–Present
Region	Initially Asia then extended to Africa and Latin America; socialist countries to nationalist countrics	Extended to Oceania, centred in Africa	concentrated in Asia and Africa; focusing the BRI alongside countries
Designer	Local architects of the recipient counties; China's state-owned design institutes	China's state-owned design institutes	China's state-owned institute; cooperated with foreign firms sometimes
Working Mode	Assigned by the Chinese government	Assigned by the Chinese government	Through tender and bid with multi-cooperation and competition
Style	Socialist modernism; Functional modernity	Conventional economic designs	Internationalized design with critical regionalism
Structure	RC frame structure	Developed RC frame structure; simple steel truss structure	Complex space structures (steel arch structure, membrane structure, cable-suspended structure, etc.) with new materials
Technology	Low-passive technology sometimes used	Low-passive technology frequently used	Newly developed materials, computerized simulation and other high technologies

The result of this study generates new perspectives into the current researches about China's foreign-aided constructions. This study fills the gap in the academic literature and enriches the contexts of architectural theories such as the exported modernism, politics and architecture, and critical regionalism. It also provides reference and guidance for future aided construction projects and yields a better understanding of China's foreign aid systems at present. Besides, this study makes a significant contribution to offer a research approach about the architectural output under complicated circumstances, and about the relationship between the influential factors and the architectural results. The author believes that it can also be applied to the study of other categories of China's foreign-aided buildings, such as other public buildings, including theatres and convention halls, etc. Although there might be some problems in China's foreign-aided stadiums, China's foreign-aided stadiums offer an alternative approach to improving social equality, and citizens' quality of life, contributing to the modernisation and urbanism of the majority of the recipient areas.

Bibliography

[1] 蔼炳根, 1987. 西萨摩亚体育设施, 西萨摩亚[J]. 世界建筑 (2): 28-29.

[2] 常威,薛求理, 2018. 浅谈我国援外体育建筑的地域性设计尝试[J]. 建筑与文化, 10(175): 241-243.

[3] 常威, 薛求理, 2019. 援外建筑创作中的文化表达——以援柬埔寨建筑实践为例[J]. 城市建筑, 9(330): 26-29.

[4] 常威, 薛求理, 贾开武, 2019. 我国援外体育建筑设计中地域文化的理解与表达——以三个援建体育场为例[J]. 建筑师, 6(202): 96-99.

[5] CSWADI, 2015. 大跨度空间结构工程实践 CSWADI 设计案例[M]. 北京:中国建筑工业出版社.

[6] 崔振亚, 张国庆,1991. 国家奥林匹克体育中心综合体育馆屋盖结构设计[J]. 建筑结构学报, (1): 24-37.

[7] 邓艳任, 2018. "一带一路"背景下中国对柬埔寨援助研究[J]. 广东农工商职业学院学报, 34(1): 18-21.

[8] 丁先昕, 1986. 西双版纳体育馆方案[J]. 建筑学报, 7: 16-18+82-83.

[9] 房志民, 1988. 我国援建巴巴多斯体育馆项目设计合同签字[J]. 国际经济合作 (6): 7.

[10] 付毅智, 2008. 标志性体育建筑与功能性体育建筑——回顾北京奥运中心区三大场馆的设计与建设[J]. 城市建筑, 11: 18-21.

[11] 葛如亮, 1957. 大型公共体育馆设计[J]. 建筑学报, 9: 33-39.

[12] 谷继坤, 2015. 中国工人"赴蒙援建"问题的历史考察(1949—1973)[J]. 中共党史研究 (4): 49-62.

[13] 顾军, 2004. 中国援助柬埔寨吴哥古迹周萨神庙保护工程[C]//砖石类文物保护技术研讨会论文集. 北京: 中国文物保护技术协会.

[14] 郭体元,1983. 我国援外体育场馆建筑[J]. 体育文化导刊, 1: 15-20.

[15] 韩晓青, 2014. 试论中巴建交及建交初期的两国关系[J]. 党史研究与教学 (6): 44-52.
[16] 韩秀春, 1983. 游泳建筑[J]. 建筑学报 (8): 56-62.
[17] 胡兴安, 魏敦山, 2010. 中国体育建筑 60 年回顾——魏敦山院士访谈[J]. 城市建筑 (11): 11-12.
[18] 黄至贤, 1987. 帐篷式体育建筑 沙特阿拉伯法赫德国际体育场[J]. 建筑结构 (3): 62.
[19] 贾倍思, 2003. 型和现代主义[M]. 北京：中国建筑工业出版社.
[20] 江宏, 2007. 坦桑尼亚国家体育场[J]. 建筑创作, 91(1): 50-55.
[21] 蒋乐思, 1984. 浅谈体育与美学的关系[J]. 体育科学 (2): 83-86.
[22] 姜伟光, 2013. 我国援外成套建筑工程项目施工质量管理研究[D]. 长沙: 中南大学.
[23] 廖杰, 2000. 可持续发展理念在体育馆设计中的运用——谈宿迁市文体综合馆的设计[J]. 安徽建筑 (6): 47-48.
[24] 刘方平, 2016. 中国援外的历史进程与现实拓展[J]. 暨南学报（哲学社会科学版）(2): 120-128.
[25] 刘方平, 2016. 中国援外战略转变探析[J]. 东北亚论坛, 3: 49-58.
[26] 刘宏伟, 2008. 大跨建筑混合结构的分类[J]. 城市建筑, 1: 29-31.
[27] 刘凯, 2019. “一带一路”视角下中国对东南亚援助研究[D]. 上海: 上海师范大学.
[28] 刘振秀, 1990. 奥林匹克体育中心游泳馆[J]. 建筑学报 (9): 20-25.
[29] 陆赐麟, 1983. 国外大型体育建筑结构的发展与趋向[J]. 建筑结构学报 (5): 71-77.
[30] 罗文昌, 2017. 援外成套建筑工程项目施工质量评价于控制研究——以援喀麦隆社会住房项目为例[D]. 沈阳: 沈阳建筑大学.
[31] 马国馨, 1991. 体育、建筑、城市——亚运会设施建设随感[J]. 建筑师, 3: 1-21.
[32] 马国馨, 2010. 体育建筑 60 年[J]. 城市建筑 (11): 6-10.
[33] 马国馨, 2019. 新中国体育建筑 70 年[M]. 北京：中国建筑工业出版社.
[34] 梅季魁, 刘德明, 姚亚维, 2002. 大跨建筑结构构思与结构选型[M]. 北京：中国建筑工业出版社.
[35] 孟晓勇, 延汝萍, 2011. 坦桑尼亚体育场膜结构施工技术[J]. 建筑技术, 42(7): 628-630.
[36] 蒙古国中国历史文化研究协会, 2019. 我们知道的和不知道的中蒙友谊[M]. 蒙古：中国历史与文化蒙古研究协会.
[37] NEVILLE PL, 1981. 建筑的艺术与技术[M]. 北京：中国建筑工业出版社.

[38] 牛海，刘科元，2016. 索膜结构体育馆发展综述[M]. 科技视界 (10): 41-42.

[39] 庞陈敏,韩煜皎,刘乃山, 2005. 最大规模的中国民间对外援助——中国民间援助印度洋海啸灾区的简要回顾[J]. 中国减灾 (3): 26-27.

[40] 平言, 2000. 点染江山六十春——记九三学社四川省委名誉主席建筑设计大师徐尚志[J]. 四川统一战线 (5): 10-11.

[41] 钱锋,余中奇, 2015. 结构建筑学——触发本体创新的建筑设计思维 [J]. 建筑师, 2: 26-32.

[42] 钱可权,赵小彭, 徐强生, 1981. 体育馆建筑设计中的几个问题[J]. 建筑学报 (4): 50-51.

[43] 饶洁, 2008. 体育场馆国际设计竞赛创作理念研究[D]. 哈尔滨: 哈尔滨工业大学.

[44] 石林,1989. 当代中国的对外经济合作[M]. 北京：中国社会科学出版社.

[45] 苏联部长会议体育运动委员会运动构筑物设计标准体育场与体育馆 H110-53[S]. 上海工业及城市建筑设计院, 译. 北京：建筑工程出版社.

[46] 苏钧, 1989. 坚持改革开放竭诚为发展中国家服务——中国成套设备进出口公司在前进[J]. 国际经济合作, 10: 29-30.

[47] 孙广勇，2012. 中柬合作、携起友谊之手[N]. 人民日报，2012-03-20.

[48] 孙一民, 1999. 基本问题的解决与思考——中山体育馆设计反思[J]. 华中建筑, 3: 59-61.

[49] 清华大学建筑学院, 1979. 国外建筑实例图集——体育建筑[M]. 北京：中国建筑工业出版社.

[50] 建设部勘察设计司, 1999. 中国建筑设计精品集锦 (5)[M]. 北京：中国建筑工业出版社.

[51] 江苏省人民政府办公厅, 1993. 对外贸易与经济合作.1993 江苏年鉴-1993[M]:333-334.

[52] 中国外交部, 1950. 外交部亚洲司关于在蒙华侨及蒙古的一般情况的报告[A]. 北京: 中国外交部档案馆, No. 106-00025-03.

[53] 中国外交部，1950. 外交部办公厅关于在蒙华侨要求回国问题致吉雅泰的函[A]. 中国外交部档案馆, No. 106-00025-03.

[54] 中华人民共和国外交部外交史研究室, 1987. 中国外交概览[M]. 北京：世界知识出版社.

[55] 中国国务院，1955. 习仲勋就国务院关于蒙古回国工人的安置问题给北京等地的电报[A]. 北京: 北京市档案馆，No.002-020-00738.

[56] 中国国务院，1956. 关于中华人民共和国给予蒙古人民共和国经济和技术援助的协定，北京市道路工程局关于城建蒙古人民共和国市内道路工程给北京市人

民委员会的请示报告[A]. 北京: 北京市档案馆，No.002-008-00162.
[57] 中国国务院，1957. 国务院关于决定从河北省动员调遣 1 000 名工人赴蒙古参加经济建设的通知[A]. 石家庄: 河北档案馆, No. 855-4-1241.
[58] 王宝峰, 2008. 体育馆建筑设计手法及发展趋势[D]. 天津: 天津大学.
[59] 王冲，2014. 中国对外援助 60 年变迁史[J]. 党政论坛, 2: 42-44.
[60] 王惠德,1990. 北京国际网球中心[J]. 建筑学报, 10: 50-51.
[61] 魏治平,初晓,陆诗亮, 张玉影, 2013. 旋·彩——大连市体育中心体育场设计[J]. 建筑学报, 10: 64-65.
[62] 王正夫, 1986. 国外体育建筑的设计思想[J]. 建筑学报 (7): 5-11.
[63] 项端祈,王峥,陈金京,1991. 国家奥林匹克体育中心综合体育馆和游泳馆的声学设计[J]. 应用声学 (1), 1-8.
[64] 萧默, 1991. 传统与现代交融的契机——国家奥林匹克体育中心建筑艺术启示录[J]. 美术 (10): 60-62+67-73.
[65] 新华网, 2018. 关于国务院机构改革方案的说明[N 10L].[63-17]. http://www.cidca.gov.cn/2018-11/13/c_129992970.htm.
[66] 许梁, 2008. 我国对外援助项目的管理研究——以老挝项目为例[D]. 昆明: 昆明理工大学.
[67] 徐尚志，2003. 在内罗毕的日日夜夜. 建筑百家回忆录续[M]. 北京：知识产权出版社，中国水利电力出版社.
[68] 许世文，2002. 环境·功能·形式·发展——杭州黄龙体育中心主体育场设计体会点滴[J]. 新建筑 (6): 43-46.
[69] 薛求理,史巍,2003. 建筑设计招投标：优选还是负累[J]. 建筑师，4: 29-32.
[70] 杨秉德, 1991. 亚运之星——石景山体育馆与北京体育学院体育馆述评[J]. 新建筑 (2): 16-18.
[71] 杨为华,1994. 援外工程的又一探索——巴巴多斯体育馆设计[J]. 新建筑, 3: 3-5.
[72] 杨永生, 2002. 中国四代建筑师[M]. 北京：中国建筑工业出版社.
[73] 尹淮,1956. 重庆市人民体育场[J]. 建筑学报 (9): 12.
[74] 由保贤,2017. 中巴友谊的结晶——援建巴基斯坦综合体育设施工程技术解析[M]. 北京：中国建筑工业出版社.
[75] 俞大伟，李勇勤，2016. 无偿与合作：我国体育对外援助方式研究[J]. 武汉体育学院学院报, 50 (6): 22-29.
[76] 俞大伟,袁雷, 2010. 我国体育对外援助的历史回顾[J]. 北京体育大学学报, 8: 39-41.
[77] 于涌泉, 2016. 中国对外援助状况研究(1949—2010)[D]. 长春: 吉林大学.

[78] 袁尧, 2010. 黎佗芬体育场馆建筑设计研究[D]. 成都: 西南交通大学.

[79] 张光恺,1990. 首都速度滑冰训练馆[J]. 建筑学报, 11: 32-33.

[80] 张婷, 1991. 国家奥林匹克体育中心游泳馆扩声系统介绍[J]. 电声技术 (3):24-27.

[81] 张耀曾,刘振秀, 郭恩章, 1984. 我国体育馆建筑的实践与问题——体育馆建筑三十五[J]. 长安大学学报 (建筑与环境科学版), 2: 53-76.

[82] 张宁秋,2008. 中哥建交援建献礼——访援哥斯达黎加国家体育场项目负责人[J]. 华中建筑, 26 (8): 258-259.

[83] 张郁惠, 2006. 中国对外援助研究[D]. 北京: 中共中央党校.

[84] 张郁慧,2012. 中国对外援助研究 1950—2010[M]. 北京:九州出版社.

[85] 赵基达， 1988. 网壳结构在大跨度体育建筑中的应用. 第四届空间结构学术交流会论文集, 144-149[C]. 北京: 中国土木工程学会桥梁及结构工程学会空间结构委员会, 中国土木工程学会.

[86] 赵瑾兮, 2016. 勘查设计企业在援外成套项目新模式下面临的机遇、挑战及对策研究[D]. 北京: 对外经贸大学.

[87] 郑祥, 2016. 气膜体育馆及其发展的建设性研究[J]. 新型建筑材料, 43(5): 84-86+94.

[88] 周方中, 1984. 探索与追求——一个游泳馆的设计构思[J]. 建筑师, 19(6): 35-44.

[89] 周方中, 1996. 浪漫与理性共生的追求——珠海市体育中心游泳馆的创作思考[J]. 新建筑, 3: 3-6.

[90] 周弘, 2010. 中国援外六十年的回顾与展望[J]. 外交评论, 27(5): 3-11.

[91] 周治良, 1983. 慕尼黑奥林匹克体育中心西德[J]. 世界建筑 (5): 23-27.

[92] 朱竞翔, 吴程辉, 何英杰, 等, 2019. 菲律宾锡亚高阿普萨拉斯酒店[J]. 世界建筑导报, 34(1): 38-41.

[93] 朱思荣, 1988. 联邦德国体育建筑结构的发展趋向. 第四届空间结构学术交流会论文集[C]. 北京: 中国土木工程学会桥梁及结构工程学会空间结构委员会, 中国土木工程学会.

[94] 朱晓明,吴杨杰， 2018. 独立与外援:柬埔寨新高棉建筑及总建筑师凡·莫利万作品研究[J]. 时代建筑, 6:131-135.

[95] 邹德侬, 2003. 中国现代建筑史[M]. 北京:机械工业出版社.

[96] 佚名, 1974. 采用鞍形悬索屋盖结构的浙江人民体育馆[J]. 建筑学报 (3): 38-43+30-46.

[97] 佚名,1994. 珠海体育中心游泳馆[J]. 建筑学报 (12): 8-9.

[98] 佚名, 1995. 东莞体育中心体育场[J]. 建筑学报 (10): 19.

[99] 佚名, 1997. 成都市体育中心体育场设计[J]. 四川建筑 (1): 31-32.

[100] 佚名, 2006. 坦桑尼亚国家体育场万向支座设计及试验分析研究[J]. 建筑结构, 36(S1): 419-421.

[101] 作者不详, 2015. 企业研究 (1): 9.

[102] ALEXANDER C, 1977. A pattern language: towns, buildings, construction[M]. New York: Oxford University Press.

[103] ALDEN C, 2007. China in Africa[M]. London: Zed Books.

[104] AMOAH L, 2016. China, architecture and Ghana's spaces: concrete signs of a soft Chinese imperium? [J]. Journal of Asian and African Studies, 51 (2): 238-255.

[105] ANONYMOUS, 1996. Architecture, 4:77.

[106] AROONPIPAT S, 2018. Governing aid from China through embedded informality: institutional response to Chinese development aid in Laos[J]. China Information, 32(1): 46-68.

[107] AURELI P V, 2011. The possibility of an absolute architecture[M]. Cambridge, Mass: MIT Print.

[108] BABACI-WILHITE Z, MACLEANS A, GEO-JAJA, et al, 2013. China's aid to Africa: competitor or alternative to the OECD aid architecture? [J]. International Journal of Social Economics, 40(8): 729-43.

[109] BARTKE W, 1989. The economic aid of the PR China to developing and socialist countries[M]. New York: K.G. Saur.

[110] BEECKMANS L, 2017. The architecture of nation-building in Africa as a development aid project: designing the capitols of Kinshasa (Congo) and Dodoma (Tanzania) in the post-independence years[J]. Progress in Planning, 122: 1-28.

[111] BERMAN M, 1988. All that is solid melts into air: the experience of modernity[M]. New York: Penguin Books. 15-36.

[112] BERNARD H R, RYAN G W, 2010. Analysing qualitative data systematic approaches[M]. London: Sage.

[113] BOGNAR B, 2000. "Surface above all? American influence on Japanese," in transactions, transgressions, transformations: American culture in Western Europe and Japan.[M]. New York: Berghahn.

[114] BRAUTIGAM D, 1998. Chinese aid and African development: exporting green revolution[M]. Basingstoke, Hampshire: New York: Macmillan; St. Martin's.

[115] BRAUTIGAM D, 2008. China's African aid-transatlantic challenges[M]. Washington, D.C.: Gernman Marshall Fund of the United State.

[116] BRAUTIGAM D, 2011a. The dragon's gift: the real story of China in Africa[M].

New York: Oxford University Press.

[117] Brautigam D, 2011b. China in Africa: seven myths [C]. The Elcano Royal Institute/ Real Instituto Elcano, Analysis of the Real Elcano Institute (ARI), 23, Madrid, Spain 8.

[118] BUSSE M, CEREN E, HENNING M, 2016. China's impact on Africa—the role of trade, FDI and aid[J]. KYKLOS, 69(2): 228-262.

[119] CANIZARO V B, 2007. Architectural regionalism: collected writings on place, identity, modernity, and tradition[M]. New York: Princeton Architectural Press.

[120] CASSIRER E, 1983. An essay on man: an introduction to a philosophy of human culture[M]. S.l.: S.n.

[121] CCCPC Party Literature Research Office, 1997. Chronicle of Zhou Enlai (1949–1976), vol Ⅰ [M]. Beijing: Central Party Literature Press.

[122] CERVER F C, 1997. 世界新建筑(体育建筑)[M]. 台北:淑馨出版社.

[123] CHANG W, XUE C Q L, 2019a. Towards international—China-aid stadiums in the developing world[J]. Frontiers of Architectural Research, 5: 604-619.

[124] CHANG W, XUE C Q L, 2020. Climate, standard and symbolization: critical regional approaches in designs of China-aided stadiums[J]. Journal of Asian Architecture and Building Engineering, 19(4): 341-353.

[125] CHANG W, XUE C Q L, DING G, 2019. Architecture of diplomacy: Chinese construction aid in Asia, 1950-1976[J/OL]. Arena Journal of Architectural Research, 16. https://ajar.arena-architecture.eu/articles/10.5334/ajar.147/.

[126] CODY J, 2003. Exporting American Architecture, 1870—2000 (planning, history and environment series) [M]. London; New York: Routledge.

[127] COHEN J L, 1995. Scenes of the world to come: European architecture and the American challenge, 1893-1960[M]. Paris, Montreal: Flammarion and the Canadian Centre for Architecture.

[128] COHEN M, 2008. How much Brunelleschi? a late medieval proportional system in the Basilica of San Lorenzo in Florence[J]. Journal of the Society of Architectural Historians, 67(1), 18-57.

[129] COPPER J F, 1979. China's foreign aid in 1978[D]. Maryland: School of Law, University of Maryland.

[130] COPPER J F, 2015. China's foreign aid and investment diplomacy: history and practice in Asia, 1950-present, vol. Ⅱ [M]. New York: Palgrave Macmillan.

[131] DAVIES P, 2007. China and the end of poverty in Africa—towards mutual benefit?

[M]. Sundyberg, Sweden: Diakonia, Alfaprint.

[132] DAY G, 2010. The project of autonomy: politics and architecture within and against capitalis, Pier Vittorio Aureli, New York: the Temple Hoyne Buell Center for the study of American architecture at Columbia University and Princeton Architectural Press, 2008." [J]. Historical Materialism, 18(4): 219-236.

[133] DING G, XUE C Q L, 2015. China's architectural aid: exporting a transformational modernism[J]. Habitat International, 47(1): 136-147.

[134] DONG Y, FAN C, 2017. The effects of China's aid and trade on its ODI in African countries[J]. Emerging Markets Review, 33, 1-18.

[135] DOWNE-WAMBOLDT B, 1992. Content analysis: method, applications, and issues[J]. Health Care for Women International, 13 (3): 313-321.

[136] DOYTCHINOV G, 2012. Pragmatism, not ideology: Bulgarian architectural exports to the "Third World" [J]. The Journal of Architecture, 17 (3): 453-473.

[137] DREHER A, FUCHS A, 2015. Rogue aid? an empirical analysis of China's aid allocation[J]. Canadian Journal of Economics/Revue Canadienne D'économique, 48(3): 988-1023.

[138] DREHER A, NUNNENKAMP P, THIELE R, 2011. Are "new" donors different? comparing the allocation of bilateral aid between Non-DAC and DAC donor countries[J]. World Development, 39(11): 1950-1968.

[139] EM-DAT, 2012. EM-DAT: The OFDA/CRED international disaster database—www.emdat.be[M]. Brussels, Belgium: Université Catholique de Louvain.

[140] FAN H, LU Z, 2012. Representing the new China and the Sovietisation of Chinese sport (1949-1962) [J]. The International Journal of the History of Sport: Communists and Champions—The Politicisation of Sport in Modern China, 29(1): 1-29.

[141] FAN H, XIONG X, 2005. Communist China: sport, politics and diplomacy[J]. The International Journal of the. History of Sport, 19(2-3): 257-276.

[142] FLOWERS B, 2018. Sport and architecture[M]. London and New York: Routledge.

[143] FOSTER V, BUTTERFIELD W, CHEN C, ET AL, 2009. Building bridges—China's growing role as infrastructure financier for sub-Saharan Africa[M]. Washington D.C.: The World Bank.

[144] FRAMPTON K, 1992. Modern architecture: a critical history[M]. London: Thames and Hudson.

[145] FRAMPTON K, 2007. Ten points on an architecture of regionalism: a provisional polemic[M]//CANIZARO V B, Architectural regionalism: collected writings on

place, identity, modernity, and tradition[M]. New York: Princeton Architectural Press, 375–385.

[146] FRASER M, KERR J, 2007. Architecture and the "special relationship" : the American influence on post-war British architecture[M]. London: Routledge.

[147] FUCHS A, 2014. Determinants of donor generosity: a survey of the aid budget literature[J]. World Development, 56: 172.

[148] GAO H, HE L, 1995. History of the People's Republic of China[M]. Beijing: China Archives Press.

[149] GLASER B, STRAUSS A, 1967. The discovery of grounded theory: strategies for qualitative research[M]. New York: Aldine de Gruyter.

[150] GOULD K, 1999. Amerian architects export collaboration as a mean to sustainablitliy[J]. AIArchitect, (November): 23.

[151] GUEST G, MAXQUEEN K, Namey E, 2012. Applied thematic analysis[M]. Thousand Oaks, CA: Sage Publications.

[152] HANNIGAN J, 1998. Fantasy city: pleasure and profit in the postmodern metropolis[M]. London, New York: Routledge.

[153] HARVEY D, 1989. The condition of postmodernity: an enquiry into the origins of cultural change[M]. Oxford: Blackwell.

[154] HENRIQUE K P, 2013. Modernity and continuity: alternatives to instant tradition in contemporary Brazilian architecture[J]. Spaces Flows: Int. J. Urban Extra. Stud, 3 (4): 103–112.

[155] HOGAN M, 1989. The marshall plan: America, Britain and the reconstruction of Western Europe, 1947-1952[M]. New York: Cambridge University Press.

[156] HOGBEN P, 2012. Architecture and arts' and the mediation of American Architecture in post-war Australia[J]. Fabrications, 1: 30.

[157] HUBBARD P, 2017. Aiding transparency: what we can learn about China ExIm Bank's concessional loans. CGD working paper 126[M]. Washington DC: Centre for Global Development.

[158] JOHN T, 2010. The pursuit of history[M]. 5th ed. UK: Pearson Education Limited.

[159] LAUREN J, 2019. The Belt and Road Initiative: what is in it for China?[J]. Asia & the Pacific Policy Studies, 6 (1): 40-58.

[160] JULIUS P, 2004. Dari Gelora Bung Karno ke Gelora Bung Karno (in Indonesian) [M]. Jakarta: Grasindo.

[161] KACEL E, 2010. This is not an American house: good sense modernism in 1950s

Turkey[M]//DUANFANG L, Third world modernism: architecture, development and identity, chapter 7. London, New York: Routledge, 165-185.

[162] KHATTAK A, KHALID I, 2017. China's One Belt One Road Initiative: towards mutual peace and development[J]. Journal of the Research Society of Pakistan, 54(1).

[163] KILAMA E, 2016. The influence of China and emerging donors aid allocation: a recipient perspective[J]. China Economic Review, 38: 76-91.

[164] KITANO N, 2018. China's foreign aid: entering a new stage[J]. Asia-Pacific Review, 25(1): 90-111.

[165] KOBAYASHI T, 2008. Evolution of China's aid policy. JBICI working paper no. 27[M]. Tokyo: Japan Bank for International Cooperation Institute.

[166] KORIN P, 1971. Thoughts on art, in: socialist realism in literature and art[M]. Moscow: Progress Publishers, 95.

[167] KRIEGER P, 2000. Learning from America: postwar Urban recovery in West Germany[M]// FEHRENBACH H, POIGER U G. Transactions, transgressions, transformations: American culture in Western Europe and Japan. New York: Berghahn.

[168] KUCKARTZ U, MCWHERTOR A, 2014. Qualitative text analysis: a guide to methods, practice & using software[M]. London and Los Angeles: SAGE.

[169] LANCASTER C, 2007. Foreign aid: diplomacy, development, domestic politics[M]. Chicago: University of Chicago Press.

[170] LEACH N, 1999. Architecture and revolution: contemporary perspectives on Central and Eastern Europe[M]. London, New York: Routledge.

[171] LEETARU K, 2010. The Scope of FBIS and BBC open-source media coverage, 1979-2008 (U)[J]. Studies in Intelligence, 54 (1): 17-37.

[172] LEFAIVRE L, TZONIS A, 2003. Critical regionalism: architecture and identity in a globalized world[M]. Munich, New York, Prestel: Architecture in Focus.

[173] LIAMPUTTONG P, EZZY D, 2005. Qualitative research methods[M]. South Melbourne: Oxford University Press.

[174] LOEFFLER J C, 1998. The architecture of diplomacy: building America's embassies[M]. New York: Princeton Architectural Press.

[175] LOGAN W, 1995. Russians on the red river: the Soviet impact on Hanoi's townscape, 1955-90[J]. Europe-Asia Studies, 47(3): 443-468.

[176] LOGAN W, 2000. Hanoi: biography of a city[M]. Seattle, WA: University of Washington Press.

[177] LU D, 2011. Introduction: architecture, modernity and identity in the third world[M]// LU D, Third world modernism: architecture, development and identity. New York: Routledge, 1-28.

[178] LUM T, FISCHER H, GOMEZ-GRANGER J, ET AL, 2009. China's foreign aid activities in Africa, Latin America, and Southeast Asia. Congressional Research Service Report for Congress[M]. Washington D.C.: Congressional Research Service.

[179] MCKAY S, VERTINSKY P, 2004. Disciplining bodies in the gymnasium: memory, monument, modernity[M]. New York: Taylor and Francis.

[180] MEHROTRA R, 2011. Architecture in India since 1990[M]. Mubai: Ostfildern; Pictor: Hatje Cantz.

[181] MENARY S, 2015. China's programme of stadium diplomacy[J]. ICSS Journal, 3(3): 2-9.

[182] MILES M, HUBERMAN A M, 1994. Qualitative data analysis [M]. Thousand Oaks, CA: Sage Publications.

[183] MILNE D, 1981. Architecture, politics and the public realm[J]. Canadian Journal of Political and Social Theory, 5: 1-2.

[184] MONSON J, 2009. Africa's freedom railway: how a Chinese development project changed lives and livelihoods in Tanzania[M]. Bloomington: Indiana University Press.

[185] NAÍM M, 2007. Missing links: rogue aid[J]. Foreign Policy, 159: 96-95.

[186] NELSON R, 2017. Locating the domestic in Vann Molyvann's National Sports Complex[J]. ABE Journal (Architecture beyond Europe), 11.

[187] NEUMAYER E, 2003. Do human rights matter in bilateral aid allocation? a quantitative analysis of 21 donor countries[J]. Social Science Quarterly, 84(3): 650–666.

[188] NIU Z, 2016. China's development and its aid presence in Africa: a critical reflection from the perspective of development anthropology[J]. Journal of Asian and African Studies, 51(2): 199-221.

[189] NORAGRIC P W, 2011. China and Africa—aid and development[M]. EDS 270 Development Aid & Politics.

[190] NUTTALL I, 2008. Kicking Off[J]. Stadia, 7.

[191] OECD, 1987. The aid programme of China[M]. Paris: Organisation for Economic Co-operation and Development.

[192] PAN Y, 2015. China's foreign assistance and its implications for the international

aid architecture[J]. China Quarterly of International Strategic Studies, 1: 283-304.

[193] PATTON M, 1990. Qualitative evaluation and research methods[M]. Beverly Hills, CA: Sage.

[194] PERERA N, TANG W, 2013. Transforming Asian cities: intellectual impasse, asianizing space, and emerging translocalities [M]. Abingdon: Oxon; New York: Routledge.

[195] PIERIS A, 2011. "Tropical" cosmopolitanism? the untoward legacy of the American style in post-independence Ceylon/Sri Lanka[J]. Singapore Journal of Tropical Geography, 32(3): 332-349.

[196] PONLOK Y, CHAMROEUN U, 2012. Phnom Penh sports complex plans laid bare[M/OL]. Phnom Penh Post. https://www.phnompenhpost.com/sport/phnom-penh-sports-complex-plans-laid-bare.

[197] RADCLIFFE-BROWN A, 2011. As quoted in Giampietro Gobo, "Ethnography" [M]//SILVERMAN D, Qualitative research: issue of theory, method and practice, 3rd ed. Log Angeles: SAGE.

[198] RAGIN C C, 1987. The comparative method: moving beyond qualitative and quantitative strategies[M]. London: University of California Press.

[199] RALEIGH C, LINKE A, HEGRE H, ET AL, 2010. Introducing ACLED: an armed conflict location and event dataset: special data feature[J]. Journal of Peace Research, 47(5): 651-660.

[200] REILLY J, 2012. A norm-taker or a norm-maker? Chinese aid in Southeast Asia[J]. Journal of Contemporary China: the Rise of China and the Regional Responses in the Asia-Pacific, 21(73): 71-91.

[201] RIFFE D, LACY S, FICO F, 2014. Analyzing media messages: using quantitative content analysis in research, Routledge communication series)[M]. 3rd ed New York: Routledge.

[202] RIHOUX B, 2003. Bridging the gap between the qualitative and quantitative worlds? a retrospective and prospective view on qualitative comparative analysis[J]. Field Methods, 15(4): 351-365.

[203] RIHOUX B, 2006. Qualitative comparative analysis (QCA) and related systematic comparative methods: recent advances and remaining challenges for social science research[J]. International Sociology, 21(5): 679-706.

[204] RING A, HENRIETTE S, KRISTIN V, 2018. Architecture and control[M]. Leiden, Boston, Brill: Architectural Intelligences.

[205] ROBIN R T, 2014. Enclaves of America: the rhetoric of american political architecture abroad, 1900-1965, Princeton Legacy Library[M]. Princeton: Princeton University Press.

[206] ROSKAM C, 2015. Non-aligned architecture: China's designs on and in Ghana and Guinea, 1955-92[J]. Architectural History, 58: 261-291.

[207] ROSS E, 2014. China's stadium diplomacy in Africa[J/OL]. Roads & Kingdoms. [2018-01-27]. https://roadsandkingdoms.com/2014/chinas-stadium-diplomacy-in-africa/

[208] ROWE P G, KUAN S, 2002. Architectural encounters with essence and form in modern China[M]. Cambridge, Mass.: MIT Press.

[209] SALEHYAN I, HENDRIX C, HAMNER J, ET AL, 2012. Social conflict in africa: a new database[J]. International Interactions: Event Data in the Study of Conflict, 38(4): 503-511.

[210] SAM M P, HUGHSON J, 2010. Sport in the city: cultural and political connections[J]. Sport in Society: Sport in the City, 13(10): 1417-1422.

[211] SCHRODT P A, DEBORAH J G, 1994. Validity assessment of a machine-coded event data set for the Middle East, 1982-1992[J]. American Journal of Political Science, 38 (3): 825-854.

[212] SCHWENKEL C, 2017. The afterlife of East German Planning in Vietnam: insurgent architecture and the remaking of urban space. "The Design Institute: building a transnational history", seminar at HKU.

[213] SCOTT J, 1998. Seeing like a state: how certain schemes to improve the human condition have failed (the Yale ISPS series)[M]. New Haven: Yale University Press.

[214] Sheard R, 2005. Stadium: architecture for the new global culture[M]. Sydney: Pesaro Publishing.

[215] SHELLMAN S, 2008. Coding disaggregated intrastate conflict: machine processing the behavior of substate actors over time and space[J]. Political Analysis, 16(4): 464–477.

[216] Siamphukdee C, 2014a. Introduction: "export architecture" and the Cold War[J]. Journal of Export Architecture: A War of the Worlds—Cold War Projects abroad. Deakin University, Deakin: 1-2.

[217] SIAMPHUKDEE C, 2014b. What have you done? typologies of export architecture[J]. Journal of Export Architecture: A War of the Worlds—Cold War Projects Abroad. Deakin University, Deakin, 6-8.

[218] SIMSON O, 1998. The gothic cathedral[M]. New Jersey: Princeton University Press.

[219] SLESSOR C, 2000. Concrete regionalism[M]. London: Thames & Hudson.

[220] SOROKINA Y, 2012. Ghost of a garden city[M]//RITTER K, Soviet modernism 1955-1991: unknown history, 179-192. Switzerland: Park Books.

[221] SORTIJAS S, 2007. Tanzania's new national stadium and the rhetoric of development[J/OL]. Ufahamu: A Journal of African Studies, 33(2-3). http://www.escholarship.org/uc/item/4z08x0ns.

[222] STANEK L, 2012a. Introduction: the second world's architecture and planning in the third world[J]. Journal of Architecture, 17(3), 299-307.

[223] STANEK L, 2012b, Miastoprojekt goes abroad: the transfer of architectural labour from socialist Poland to Iraq (1958-1989)[J]. Journal of Architecture, 17(3), 361-386.

[224] STANEK L, 2015a. Mobilities of architecture in the late Cold War: from socialist Poland to Kuwait, and back[J]. International Journal of Islamic Architecture, 4(2), 365-398.

[225] STANEK L, 2015b. Architects from socialist countries in Ghana (1957-1967): modern architecture and mondialisation[J]. Journal of the Society of Architectural Historians, 74 (4), 416-442.

[226] STANEK L, 2019. Architecture in global socialism: Eastern Europe, West Africa, and the Middle East in the Cold War[M]. New York: Princeton University Press.

[227] STANLEY D, 1987. South Pacific handbook[J]. The Journal of Polynesian Society.

[228] STRANGE A, DREHER A, FUCHS A, ET AL, 2017. Tracking underreported financial flows: China's development finance and the aid-conflict nexus revisited.[J] Journal of Conflict Resolution, 61(5), 935-963.

[229] STRANGE A, O'DONNELL B, GAMBOA D, ET AL, 2013. AidData's media-based data collection methodology[M]. Williamsburg, VA: AidData.

[230] STRAUSS A, CORBIN J, 1998. Basics of qualitative research: techniques and procedures for developing grounded theory[M]. Thousand Oaks, CA: Sage Publications, Inc.

[231] TAYLOR I, 2006. China and Africa: engagement and compromise[M]. New York: Routledge.

[232] TAYLOR I, 2010. China's new role in Africa. Boulder[M]. Colorado: Lynne Rienner Pub.

[233] THE AFRICA REPORT, 2011. China and the Chinese stadium diplomacy[OL]. http://www.playthegame.org/fileadmin/documents/World_Stadium_Index_8_Chinese_stadium_diplomacy.pdf.

[234] THE MINISTRY OF COMMERCE OF CHINA., 2004-2009. China commerce yearbook[M]. Beijing: China Commerce and Trade Press.

[235] THE MINISTRY OF FOREIGN AFFAIRS OF CHINA, 1958. Collection of the treaties of the People's Republic of China, vol 5[M]. Beijing: Law Press.

[236] THE MINISTRY OF FOREIGN AFFAIRS OF CHINA, 1982. Collection of the treaties of the People's Republic of China, vol 22[M]. Beijing: World Knowledge Publishing House.

[237] THE STATE SPORTS COMMISSION, 1958. Proposal on the ten-year planning of sports venues (draft).

[238] THE STATE COUNCIL OF CHINA, 2011. White paper on China's foreign aid. Beijing.

[239] THE STATE COUNCIL OF CHINA, 2014. White paper on China's foreign aid. Beijing.

[240] THE SWEDISH ASSOCIATION OF ARCHITECTS, 2009. Architecture and politics—an architectural policy for Sweden, 2010-2015[M]. Swedish: Swedish Association of Architects.

[241] THOMPSON J B, 1990. Ideology and modern culture: critical social theory in the era of mass communication[M]. Cambridge: Polity.

[242] Trumpbour R C, 2006. The new cathedrals: politics and media in the history of stadium construction[M]. New York: Syracuse University Press.

[243] TZONIS A, LEFAIVRE L, 2003. Critical regionalism: architecture and identity in a globalized world[M]. New York: Prestel.

[244] UDUKU O, 2006. Modernist architecture and "the Tropical" in the West Africa: the tropical architecture movement in West Africa 1948-1970[J]. Habitat International, 30(6): 396-411.

[245] VERNON P, 2012. Shopping Towns Australia[J]. Fabrications: The Journal of the Society of Architectural Historians Australia & New Zealand, 22(1): 102-121.

[246] VORAJEE I, 2019. National Stadium Marks Construction Milestone. Khmer Times[OL].[2019-08-23]. https://www.khmertimeskh.com/603699/national-stadium-marks-construction-milestone/.

[247] WANG D, GROAT N L, 2013. Architectural research methods[M]. Toronto: John

Wiley & Sons, Incorporated.

[248] WANG T, 1999. History of diplomacy of the People's Republic of China (1970-1978), vol. 3[M]. Maharashtra: World Knowledge Publishing House.

[249] WHARTON A J, 2011. Building the Cold War—Hilton international hotels and modern architecture[M]. Chicago: University of Chicago Press.

[250] WILL R, 2011. China's stadium diplomacy[J]. World Policy Journal, 29(2): 36-43.

[251] WRIGHT C. G, 2008. Global ambition and local knowledge[M]//ISENSTADT S, RIZVI, K, Modernism and the Middle East, 221-254[M]. Seattle: University of Washington Press.

[252] XUE C Q L, 2006. Building a revolution: Chinese architecture since 1980[M]. Hong Kong: Hong Kong University Press.

[253] XUE C Q L, DING G, CHANG W, ET AL, 2019. Architecture of "stadium diplomacy"—China-aid Sport Buildings in Africa[J]. Habitat International, 90: 1-11.

[254] YEANG K，1989. Tropical urban regionalism: building in a South-East Asian city[M]. Singapore: Butterworth-Heinemann.

[255] YIN R, 2014. Case study research: design and methods[M]. 5th ed. Los Angeles: SAGE.

[256] YONAMINE J E, SCHRODT P A, 2011. A guide to event data: past, present, and future[J]. All Azimuth, 2(2): 5-22

[257] ZHANG D，SMITH G, 2017. China's foreign aid system: structure, agencies, and identities[J]. Third World Quarterly, 38(10): 2330-2346.

[258] ZHOU H, XIONG H, 2013. China's foreign aid: 60 years in retrospect[M]. Beijng: Social Sciences Academic Press.

[259] ZHOU P, 1990. Building sport facilities for foreign countries[M]//The Department of policy, Sports Ministry, 40 years' achievements of Chinese sports, Beijing: People's Sport.

[260] ZHU J, 2009. Architecture of modern China: a historical critique[M]. London, New York: Routledge.

[261] ZUNZ O, 1998. Why the American century? [M]. Chicago: University of Chicago Press.

APPENDIXES

Appendix 1 Glossary of Terms

English	Chinese	Abbreviation
Agency for International Economic Cooperation	国际经济合作事务局	AIECO
Beijing Institution of Architectural Design	北京建筑设计研究院	BIAD
China Architecture Design & Research Group	中国建筑设计研究院	CADRG
China International Development Cooperation Agency	中国国际发展合作属	CIDCA
China IPPR International Engineering Co., LTD	中国中元国际有限公司	IPPR
China Sports Industry Group	中体国际集团	CSIG
Central South Architectural Design Institute	中南建筑设计院	CSADI
Complete Plant Import & Export Corporation	成套设备进出口公司	COMPLANT
Development Assistance Committee	发展援助委员会	DAC
Export-Import Bank of China	中国进出口银行	EIBC
Games of the Newly Emerging Forces	新兴力量运动会	GANEFO
Gross Domestic Product	国内生产总值	GDP
Less Developed Countries	欠发达地区	LDC
Ministry of Commerce of the People's Republic of China	中国商务部	MOFCOM
Ministry of Economic Relations with Foreign Countries	对外经济联络部	MERFC
Ministry of Foreign Economic and Trade	外经贸部	MFET
Official Development Assistance	官方发展援助	ODA
The Silk Road Economic Belt and the 21st Century Maritime Silk Road; One Belt and One Road Initiative	"一带一路"倡议	BRI
Organization for Economic Co-operation and Development	经济合作发展组织	OECD

Continued

English	Chinese	Abbreviation
Shanghai Construction Group	上海建工集团	SCC
Shanghai Institute Architectural Design & Research Co., LTD	上海建筑设计研究院	SIADR
Shanghai Xian Dai Architectural Design (Group)	上海现代设计集团	SXDADG

Appendix 2　Basic Information of China's Foreign-aided Stadium Projects from the 1950s to 21st Century (Database of the Study)

No.	Year	Conti-nent	Country	Project Name	Design Institute	Capacity (seats)	Photo
Outdoor Stadium							
1	1956	Asia	Nepal	Dasarath Rangasala Stadium	—	25,000	
2	1958	Asia	Mongolia	National Olympic Stadium	BIADI	12,500	
3	1958	Asia	Vietnam	National Stadi-um in Hanoi	—	10,000	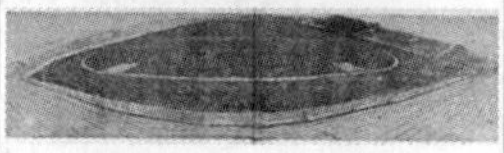
4	1963	Asia	Indonesia	Gelora Bung Karno Stadium (National Stadium)	—	110,000	
5	1968	Africa	Tanzania	Zanzibar Amaan Stadium	BIADI	10,000	
6	1973	Africa	Ethiopia	Abebe Bikila Stadium	BIADI	30,000	
7	1974	Africa	Uganda	National Stadium	BIADI	40,000	

Continued

No.	Year	Conti-nent	Country	Project Name	Design Institute	Capacity (seats)	Photo
8	1975	Africa	Sierra Leone	Siaka Stevens Stadium	Zhejiang Industrial Architectural Design Institute (ZIADI)	30,000	
9	1976	Asia	Pakistan	Main Stadium of Pakistan Sports Complex	BIADI	50,000	
10	1977	Africa	Somalia	Mogadishu Stadium	Guangxi comprehensive architectural design institute	30,000	
11	1977	Africa	Benin	Friendship Stadium (Cotonou Sports Comple)	Shanghai Indus-trial Architec-tural Design Institute (SIADI)	30,000	
12	1977	Africa	Senegal	Stade de l'Amitié (Stadium of Friendship) (Renamed) Stadc Léopold Senghor	Guangdong City Architec-tural Design Institute	60,000	
13	1980	Africa	Maurita-nia	Stade Olym-pique	East China Industrial Architectural Design Institute (Renamed as ARCPLUS)	10,000	
14	1980	Africa	Morocco	Stade Moulay Abdallah (in Labatt Sports Complex)	—	65,000	
15	1980	Africa	Gambia	Independence Stadium	—	17,000	

Continued

No.	Year	Conti-nent	Country	Project Name	Design Institute	Capacity (seats)	Photo
16	1981	Africa	Burkina Faso	August 4 Stadium	—	35,000	
17	1983	Africa	Niger	Seyni Kountche Stadium	—	30,000	
18	1984	Africa	Zimba-bwe	National Sports Stadium	Gansu Archi-tecture & Engineering Co., Ltd.	60,000	
19	1984	Africa	Senegal	One Stadium	—	50,000	
20	1985	Africa	Liberia	Samuel Kanyon Doe Sports Stadium (SKD Stadium)	—	35,000	
21	1986	Africa	Guinea-Bissau	One Stadium	—	15,000	—
22	1986	Africa	Djibouti	El Hadj Hassan Gouled Aptidon Stadium	—	10,000	
23	1987	Africa	Kenya	Moi Interna-tional Sports Complex Stadium	China Southeast Architectural Design and Research Institute.	60,000	
24	1988	Africa	Rwanda	Stade Ama-horo (Peace Stadium)	Design Institute of the Ministry of Railway (Renamed as Railway Engineering Consulting Group Co., Ltd.)	20,000	

Continued

No.	Year	Conti-nent	Country	Project Name	Design Institute	Capacity (seats)	Photo
25	1988	Africa	DR Congo	Stade des Martyrs (National Stadium)	China South-west Architectural Design and Research Institute.	80,000	
26	1989	Africa	Senegal	Stade Mawade Wade	—	8,000	
27	1989	Africa	Mauritius	Stade Anjalay	China Sports International Co., Ltd.	18,000	
28	1989	Oceania	Papua New Guinea	Sir John Guise Stadium (National Sports Center)	—	30,000	
29	1990	Africa	Central African Republic	One Stadium	—	30,000	—
30	1991	Africa	The Republic of Burundi	Onc Stadium	—	20,000	—
31	1996	Africa	Congo	Massemba-Débat Stadium	—	17,500	
32	1996	Africa	Uganda	One Stadium	—	40,000	—
33	1997	Africa	Mali	Stade du 26 Mars (National Stadium)	China Overseas Engineering Group Co., Ltd.	55,000	

Continued

No.	Year	Conti-nent	Country	Project Name	Design Institute	Capacity (seats)	Photo
34	1999	Africa	Togo	Stade de Kegue (National Stadium)	Beijing Institute of Architectural Design	30,000	
35	1999	Latin America	Grenada	National Cricket Stadium	China International Engineering Design & Consult Co., Ltd.	20,000	
36	2000	Latin America	Saint Lucia	Beausejour Stadium (Darren Sammy Cricket Ground)	—	15,000	
37	2000-2002	Asia	Vietnam	My Dinh National Stadium	Shanghai Xian Dai Architectural Design (Group) Co. Ltd. (SXDAD) With Hanoi International Group	40,000	
38	2002	Africa	Central African Republic	Barthelemy Boganda Stadium (National Stadium)	—	20,000	
39	2003	Africa	Tanzania	Benjamin Mkapa Stadium (Tanzania National Stadium)	BIAD	60,000	
40-42	2004	Africa	Djibouti	Three Small Stadiums	—	—	—
43	2004	Africa	Equatorial Guinea	Estadio de Bata (Bata Stadium)	—	35,700	

Continued

No.	Year	Conti-nent	Country	Project Name	Design Institute	Capacity (seats)	Photo
44	2004	Africa	Guinea	Stade de l'Unité (National Stadium)	CSADI	50,000	
45	2005	Latin America	Bahamas	Thomas A. Robinson National Stadium	—	15,000	
46	2005	Latin America	Jamaica	Sabina Park (Cricket Ground)	HOK & SXDAD	15,600	
47	2005	Latin America	Antigua & Barbuda	Sir Vivian Richards Stadium	HOK & Hunan Architectural Design Institute Ltd.	10,000	
48	2005	Latin America	Dominica	Stadium of Windsor Park (National Cricket Stadium)	Wuhan Architectural Design Institute (renamed as GIADR①)	12,000	
49	2006	Africa	Equatori-al Guinea	Estadio de Malabo (Malabo Stadium)	—	15,250	
50-52	2006	Africa	Mali	Three small stadiums	—	—	—
53	2007	Africa	Zambia	Levy Mwana-wasa Sports Stadium	BIAD	49,800	
54	2007	Latin America	Grenada	National Stadium	China Interna-tional Engineer-ing Design & Consult Co., Ltd. (CIEDC)	10,000	

① General Institute of Architectural Design and Research Co., Ltd.

Continued

No.	Year	Conti-nent	Country	Project Name	Design Institute	Capacity (seats)	Photo
55	2008	Asia	Laos	Laos National Stadium	CCDI	2,5000	
56	2008	Africa	Mozam-bique	Estádio Nacional do Zimpeto (new National Stadium)	Wuhan Architectural Design Institute (GIADR)	—	
57	2009	Oceania	Papua New Guinea	Wewak Sports Stadium	Hunan Architectural Design Institute Ltd.	—	
58	2009	Africa	Gabon	Stade d'An-gondjé (National Stadium)	Shanghai Construction Group	40,000	
59	2009	Latin America	Costa Rica	Estadio Nacional de Costa Rica (National Stadium)	CSADI	35,000	
60	2011	Africa	Zambia	Gabon Disaster Heroes National Stadium	BIAD	50,000	
61	2011	Africa	Congo	Stade de Djambala	—	—	
62	2011	Africa	Congo	Ewo Stadium	—	—	—
63	2012	Africa	Senegal	Stade Munici-pal de Mbour	—	5,000	
64	2012	Africa	Senegal	Stade Mame Massene Sene De Fatick	—	—	—

Continued

No.	Year	Conti-nent	Country	Project Name	Design Institute	Capacity (seats)	Photo
65	2012	Africa	Cape Verde	Stadium of Cape Verde	CIEDC	10,000	
66	2012	Africa	Congo	Brazzaville Stadium (in Brazzaville Sports Complex)	CCDI	60,000	
67	2012	Africa	Ghana	Cape Coast Sports Stadium	IPPR	15,000	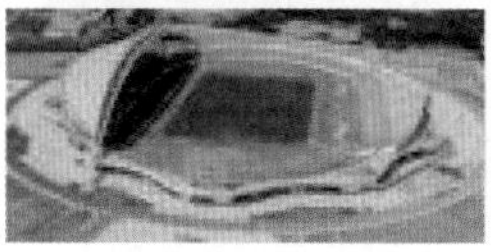
68	2013	Africa	Malawi	Bingu National Stadium	BIAD	42,900	
69	2014	Africa	Sierra Leone	Bo Stadium	—	4,000	
70	2014-2017	Africa	Cote D'lvoire	Olympic Stadium in Abidjan	BIAD	60,000	
71	2015	Africa	Comoros	Moroni Stadium	IPPR	10,000	
72	2015	Asia	Cambo-dia	National Stadium	IPPR	60,000	
73	2016	Africa	Gabon	Stade de Port-Gentil	China State Construction Engineering	20,000	

Continued

No.	Year	Conti-nent	Country	Project Name	Design Institute	Capacity (seats)	Photo
74	2016	Africa	Gabon	Stade d'Oyem	SIADR	20,031	
75	2017	Africa	Ethiopia	Addis Ababa National Stadium	MH Engineer-ing PLC	60,000	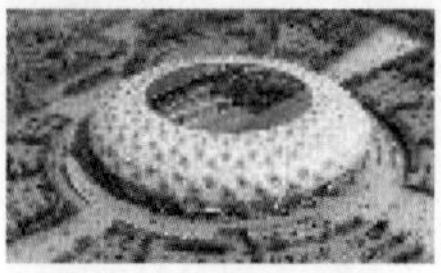
76	2017	Africa	Senegal	National Wrestling Stadium	IPPR	20,000	
77	2018	Africa	Mauritius	Stadium of Centre Culturel et Sportif (multi-purpose sports complex at Saint Pierre)	BIAD	15,000	
78	2019	Africa	Chad	Stade de N' Djamena (Stadium of Djamena)	SIADR	30,000	
79	2019	Europe	Republic of Belarus	National football stadium	BIAD	33,000	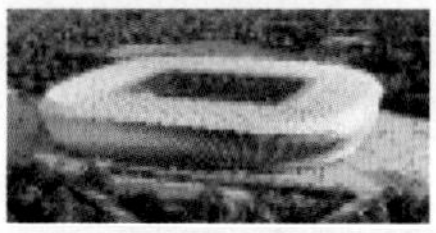
Indoor Stadium							
1	1966	Asia	Cambo-dia	Indoor Stadium (with an internation-al village) of the Olympic Sports Complex of Cambodia	BIADI	8,000	
2	1975	Asia	Syria	Damascus Tishreen Stadi-um (indoor)	BIADI	7,141	

Continued

No.	Year	Conti-nent	Country	Project Name	Design Institute	Capacity (seats)	Photo
3	1976	Asia	Pakistan	Indoor Stadium of Pakistan Sports Complex	BIADI	10,000	
4	1977	Africa	Benin	Friendship Indoor Stadium (Cotonou Sports Comple)	SIADI	5,000	
5	1980	Oceania	Samoa	Indoor stadium in Apia Park	Jiangsu Provincial Architectural D & R Institute Ltd.	1,000	
6	1980	Africa	Morocco	Indoor stadium (in Labatt Sports Complex)	—	8,000	—
7	1983	Africa	Niger	Seyni Kountche Indoor Stadium	—	3000	
8	1985	Asia	Myanmar	Thuwunna Indoor Stadium	—	10,000	
9	1985	Africa	Tunisia	Natatorium (El Menzah Youth Sports Culture Center)	—	—	
10	1986	Asia	Yemen	National Indoor Stadium	Yunan Con-struction Engineering Co. Lt.	—	

Continued

No.	Year	Conti-nent	Country	Project Name	Design Institute	Capacity (seats)	Photo
11	1986	Latin America	Surinam	Anthony Nesty Sporthal	Shanghai Civil Architectural Design Institute (Renamed as SIADR)	3,000	
12	1987	Africa	Kenya	Moi International Sports Center Gymnasium	China Southeast Architectural Design and Research Institute.	5,000	
13	1992	Latin America	Barbados	Garfield Sobers Indoor Stadium	Architectural Design and Research Institute of South East University	5,000	
14	1994	Africa	The Arab Republic of Egypt	One Stadium	—	10,000	—
15	1994	Africa	Madagascar	Tananarive Indoor Stadium	BIAD	5,000	
16	1996	Africa	Sao Tome	One Stadium		10,000	
17	2002	Oceania	Fiji	Laucala Bay Gymnasium	China Huashi Group Co., Ltd.	3,200	
18-20	2004	Africa	Morocco	Three Small Natatoriums	CADRG	—	
21	2004	Oceania	Samoa	Samoa Aquatic Center	Realway Engineering Consulting Group Co., Ltd.	—	

Continued

No.	Year	Conti-nent	Country	Project Name	Design Institute	Capacity (seats)	Photo
22	2007	Africa	Camer-oon	Sports Palace of Yaoundé	—	5,400	
23	2007	Asia	Mongolia	Buyant Ukhaa Sport Palace (National Indoor Stadium)	CIEDC	5,045	
24	2008	Asia	Laos	One Natatori-um and Two Indoor Stadiums of the National Sports Park	CCDI	—	
25	2010	Asia	Sri Lanka	Mahinda Rajapaksa International Cricket Stadium	SWA with SXDAD	35,000	
26	2012	Africa	Congo	Brazzaville Indoor Stadium (in Brazzaville Sports Complex)	CCDI	10,000	
27	2012	Africa	Congo	Brazzaville Natatorium (in Brazzaville Sports Complex)	CCDI	2,000	
28	2017	Africa	Tunisia	Natatorium (Ben Arous Youth Sports and Culture Center)	SIADR	—	

Continued

No.	Year	Conti-nent	Country	Project Name	Design Institute	Capacity (seats)	Photo
29	2017	Africa	Gabon	Libreville Indoor Stadium (for handball games)	—	5,477	
30	2017	Africa	Mauritius	Natatorium of Centre Culturel et Sportif	BIAD	1,100	—
31	2018	Africa	Algeria	Youth Sports Center	IPPR		
32	2019	Europe	Republic of Belarus	International Standard Natatorium	BIAD	6,000	

Appendix 3 List of Maintenance and Renovation/Upgrading Projects of China's Foreign-aided Stadiums

No.	Year	Continent	Country	Project	Design Institution/ Construction Contractor
1	1999	Asia	Nepal	Upgrading project of Dasarath Rangasala Stadium	China Sports International Co., Ltd.
2	2002	Africa	Congo	Maintenance of Massemba-Débat Stadium	Weihai International Economic & Technical Cooperative Co., Ltd(WEITC)
3	2004	Africa	Rwanda	Renovation of Stade Amahoro	—
4	2005	Latin America	Barbados	Renovation of Garfield Sobers Gymnasium	—
5	2005	Asian	Nepal	Upgrading Project of Dasarath Rangasala Stadium	China Sports International Co., Ltd.
6	2007	Africa	Liberia	Renovation of Samuel Kanyon Doe Sports Stadium	—

Continued

No.	Year	Continent	Country	Project	Design Institution/ Construction Contractor
7	2007	Oceania	Samoa	Maintenance and Expension of Sports Facilities in Apia Park	China National Qingjian Construction Co., Ltd. (CNQC)
8	2009	Africa	Senegal	Maintenance of Two Stadiums in Senegal	Jiangsu Provincial Construction Group
9	2010	Africa	Tanzania	Maintenance of Zanzibar Amaan Stadium	China Friendship Development International Engineering Design & Construction Co., Ltd. / Fujian Construction and Engineering Corporation
10	2011	Africa	Seychelles	Maintenance of National Swimming Pool	China Sports International Co., Ltd.
11	2012	Africa	Kenya	Renovation of Moi International Sports Center	SINOPEC
12	2012	Asia	Nepal	Dashrath Stadium Maintenance Technical Cooperation Project	China Sports International Co., Ltd.
13	2013	Africa	Senegal	Maintenance of Stade Municipal de Mbour	—
14	2014	Africa	Kenya	Renovation of Moi International Sports Center	China Southwest Architectural Design and Research Institute Co., Ltd
15	2014	Africa	Djibouti	Renovation of El Hadj Hassan Gouled Aptidon Stadium	—
16	2014	Africa	Senegal	Maintenance of Stade Mawade Wade	Zhengtai Group
17	2015	Oceania	Samoa	Maintenance of sports facilities and natatorium in Apia Park	—
18	2015	Africa	Guinea	Mantenance of Stade de l'Unité (National Stadium)	Shanghai Construction Group
19	2016	Africa	Benin	Maitenance of Friendship Stadium	Jiangxi Construction Group
20	2016	Africa	Sierra Leone	Maintenance of Bo Stadium	—
21	2017	Africa	Liberia	Renovation of Samuel Kanyon Doe Sports Stadium	HBCG International Engineering
22	2017	Latin America	Barbados	Renovation of Garfield Sobers Gymnasium	Shanxi Construction Engineering Group Co., Ltd.
23	2017	Africa	Ethiopia	Maintenance of Abebe Bikila Stadium	
24	2017	Africa	Togo	Maintenance of Stade de Kegue (National Stadium)	SIPPR Engineering Group Co., Ltd./China Geo-Engineering Corporation

Continued

No.	Year	Continent	Country	Project	Design Institution/ Construction Contractor
25	2018	Africa	Mauritania	Maintenance of Stade Olympique	Shanghai Construction Group
26	2019	Asia	Myanmar	Maintenance and Upgrading of Thuwunna Indoor Stadium	Shanghai Construction Group